From Another Kingdom
The Amazing World of Fungi

Text, 2010 ©:

Martyn Ainsworth | Lynne Boddy | Peter Crittenden
Paul Dyer | Harry C. Evans | Stephan Helfer
Patrick Hickey | Ljerka Jamnický | Heather Kiernan
Naresh Magan | Jevgenia Milne | David Minter
Andy Overall | Nick D. Read | Mike Richardson
Geoff Robson | Gordon Rutter | Milton Wainwright
Roy Watling | Martin Wishart | John Wright

Edited by:
Lynne Boddy
Max Coleman

With generous help and support from:

British Mycological Society promoting fungal science   The Scottish Government   wellcome trust

First published by the

 Royal Botanic Garden Edinburgh

20A Inverleith Row, Edinburgh EH3 5LR, UK
ISBN: 978-1-906129-67-5 Paperback edition

Note: The Royal Botanic Garden Edinburgh (RBGE) confirms that all addresses, contact details and URLs of websites are accurate at the time of going to press. RBGE cannot be responsible for details which become out of date after publication.

The views expressed are not necessarily those of the Royal Botanic Garden Edinburgh, the British Mycological Society, or the editors.

Before eating mushrooms be aware of the cautionary notes in Chapters 8 and 9 and the Recipes section.
Failure to do so is not the responsibility of the publishers, the British Mycological Society or the editors.

Every effort has been made to trace holders of copyright in text and illustrations.
Should there be any inadvertent omissions or errors the publishers will be pleased to correct them for future editions.

Designed by RBGE/Caroline Muir MCSD.
Printed by Scotprint, Scotland.

FSC   **Mixed Sources**
Product group from well-managed forests and other controlled sources
www.fsc.org Cert no. TT-COC-002217
© 1996 Forest Stewardship Council

Cover images: Main photo: *Pleurotus djamor*. Image: © Ray and Elma Kearney. Inset photos: *Calocera viscosa* (left). Image: © Peter Clarke. Bird's-nest fungus (*Cyathus striatus*) (centre); eye-lash fungus (*Scutellinia scutellata*) (right). Images: © Chris Jeffree.

2

Shaggy parasol
(*Chlorophyllum rhacodes*).
Image: © Peter Clarke.

Contents

Lactarius tabidus.
Image: © Peter Clarke.

Foreword

From Another Kingdom: The Amazing World of Fungi is aptly titled, the fungi are indeed a distinct branch in the tree of life, closer to animals than to plants and with a story so remarkable that it could have come from another world. Fungi are at once as familiar as the mushrooms on our plate and yet alien enough to trap microscopic worms in the soil or to transform caterpillars into exotic growths known in China as "summer plant – winter worm" and prized for their medicinal properties. Most of us know something about mushrooms but until the arrival of this book the full story of the fungi has been largely unknown outside the realm of the professional biologist. What makes this book so special is that it brings together some of the world's most authoritative scientists to create a beautifully illustrated, enthralling and highly readable account of the world of fungi. It is greatly to their credit that the authors have used clear and straightforward writing to tackle the full complexity of their subject, from evolutionary reconstructions based on DNA analysis to the specialised terminology needed to describe microscopic features, without compromise.

The result is a book that opens up the extraordinary diversity of fungi and explains how they impact on our daily lives and the workings of our planet. Although more than 100,000 distinct species of fungi have so far been recognised and named this is thought to be just a small fraction of the total number of species in the world. Estimates suggest that this may be as high as 1.5 million species – a figure that emphasises just how much more research is needed given the importance of the largely invisible work of fungi in nature. One essential role they play is to assist in the transfer of nutrients into the root systems of plants including many of the world's most important forest trees. Unlike plants, which can harvest the energy of the sun, fungi derive their energy from other living things, largely through the vital process of decomposition.

Left: Professor Stephen Blackmore FRSE, Regius Keeper, Royal Botanic Garden Edinburgh. Image: © RBGE/Lynsey Wilson.

Without the recycling role of decomposition the cycles of life on Earth would be incomplete. Other fungi are symbiotic in lifestyle, combining with algae to form lichens, or are parasitic, deriving their nutrition from living plants and animals, including humans. On balance, however, there is no doubt that we benefit much more than we suffer from fungi. A number of species are highly prized in cuisines around the world, some provide us with life saving medicines and others are essential for making bread and the fermentation of beer, wine and spirits. The book ends with a timely warning that fungi, in common with the other kingdoms of life, are threatened by global environmental change, especially habitat loss and climate change. Hopefully this book will open our eyes to the importance of fungi and help to inspire a new generation to discover the rewards of studying them.

Stephen Blackmore FRSE
Regius Keeper
Royal Botanic Garden Edinburgh

6

Main photos, clockwise from this page:
Mycelial cords.
Image: © Chris Jeffree.
Stereum hirsutum.
Image: © Peter Clarke.
Xanthoria species on wood.
Image: © RBGE/Vlasta Jamnický.

Facing page inset photos:
Fungal hyphae (left);
mycelial cords of
Armillaria species (right).
Images: © Patrick Hickey.

Introduction

Introducing the Fungi

Nick D. Read & Lynne Boddy

Introducing the Fungi

Nick D. Read & Lynne Boddy

Fungi are a unique and extraordinary group of organisms. They play many important roles in our everyday lives and are essential for the functioning of life on our planet. Although fungi have been lumped together with plants in the past, they are only very distantly related to them. In fact, fungi are more closely related to animals, including humans, than they are to plants. Fungi are so distinct from plants and animals that they represent an entire kingdom of organisms in their own right, and a large and diverse kingdom at that.

It has been estimated that there are around 1.5 million species of fungi, but only about 100,000 have been named so far. Recent results based on variation at the molecular level (DNA and proteins in particular) suggest that even this large figure may turn out to be a gross underestimate of the number of fungal species. The huge number of different fungi is truly remarkable when you consider that worldwide there are only about 400,000 species of flowering plants and 90% of these have been named.

Even though fungi are so numerous, their significance can be overlooked as they often remain hidden. Knowing where to look and what to look for is the first step into the fascinating and beautiful – and at times bizarre – world of fungi. This book introduces their enormous diversity and their great importance in sustaining life on Earth. It also focuses on the many ways in which fungi touch human lives, from cultural associations to their often unappreciated role in everyday life. From a human perspective fungi can kill and cure us; they can also poison and feed us. It is not surprising that there are mixed perceptions of fungi. Here the focus is on the huge debt of gratitude we all owe to fungi.

Right: Although fungi have often been lumped with plants, they are in fact not plants or animals, but a separate kingdom in their own right. Fungi are, however, closer to animals than to plants in the tree of life. In this diagram of the tree of life the groups of organisms indicated in red used to be thought of as fungi, but three of them are now known not to be closely related to the fungi.

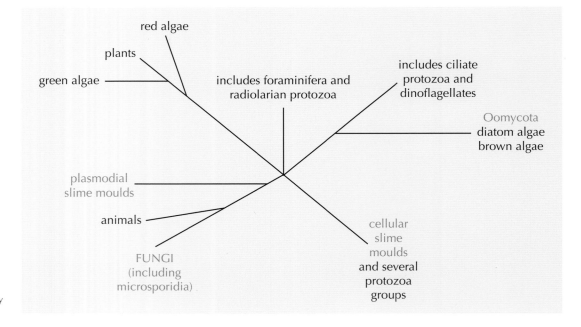

Left: Brackets, like the maze gill (*Daedalea quercina*), are one of the forms of sexual reproductive fruit bodies produced by basidiomycete fungi. They form spores that are the equivalent of the seeds of flowering plants. Image: © Stuart Skeates.

What makes a fungus?

When thinking of fungi, the first thing that comes to mind is mushrooms, toadstools (these terms can be used interchangeably) and possibly bracket fungi. However, these structures are the large sex organs of only an extremely small proportion of species within the Kingdom Fungi as a whole, and they belong to what are often loosely termed 'macrofungi'. These fungal sex organs are equivalent to the flowers/fruits of flowering plants. Even for the fungi that produce these sometimes beautiful and exotic structures, they are just 'the tip of the iceberg' of the whole fungus, its main body usually being largely hidden in whatever it is growing on. Unlike the ephemeral sex organs, known as fruit bodies, the hidden subterranean part of the fungus lives not just for a few days but for several months and often for many years.

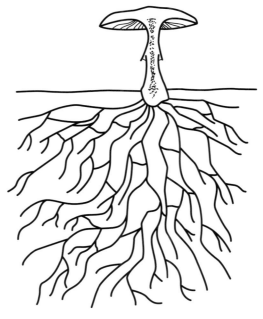

Far left: *Clitocybe nebularis*, like the bracket above, is also a basidiomycete. Its fruit body is, however, rather different and takes the commonly seen form of a stalk terminated with a cap. Such structures are often referred to as mushrooms or toadstools, but these terms are interchangeable. Image: © Peter Clarke.

Left: Fruit bodies are just 'the tip of the iceberg'; the main part of most fungi that obtains the fungus' food is the mycelium that is hidden in whatever the fungus is growing on. Illustration: © Lynne Boddy.

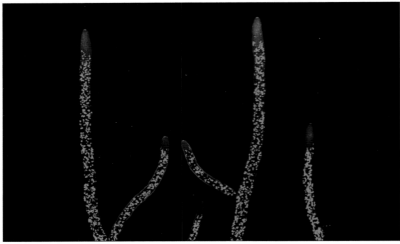

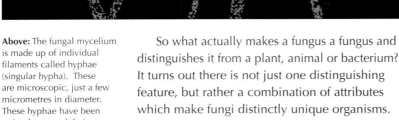

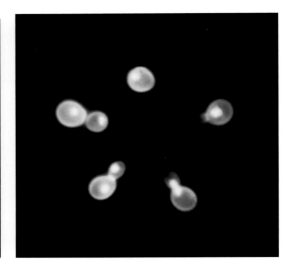

Above: The fungal mycelium is made up of individual filaments called hyphae (singular hypha). These are microscopic, just a few micrometres in diameter. These hyphae have been stained to reveal their internal structure. Image: © Patrick Hickey and Nick Read.

Above right: The other body form that fungi can take is that of yeast. Yeasts are single-celled and bud to form daughter cells. Here the yeast *Candida albicans* has been stained with a blue dye and budding is clearly visible. Image: © Neil Gow.

So what actually makes a fungus a fungus and distinguishes it from a plant, animal or bacterium? It turns out there is not just one distinguishing feature, but rather a combination of attributes which make fungi distinctly unique organisms.

The first characteristic feature is that the main feeding body of most fungi either is made of microscopic, filamentous tubes called hyphae, collectively termed a mycelium, or the fungus exists as individual yeast cells. There are also many fungi that can switch between having hyphal or yeast growth forms.

The second characteristic is that nearly all fungi do not move – a feature they share with plants. Like plants, fungi achieve 'movement' via growth, as is amply demonstrated by time-lapse photography. It is their rigid cell walls that prevent fungi from moving in the rapid

responsive way animals can. Unlike plants, in which cell walls consist largely of cellulose, most fungal cell walls contain chitin. Chitin is a tough compound, and is also found in the external skeletons of insects and other arthropods such as crustaceans.

The third characteristic of fungi is that, unlike plants, they cannot manufacture food using simple compounds and energy from sunlight through photosynthesis. Instead, they obtain their energy from the breakdown and digestion of organic matter, a feature they share with animals. Fungi feed by absorbing soluble molecules derived from the wide variety of materials they grow on and in. Small molecules such as simple sugars can be absorbed directly, but larger molecules, such as plant cellulose, must first be broken down into smaller ones.

Right: Hyphae often aggregate to form cords and these thicker structures are clearly visible with the naked eye. Image: © Chris Jeffree.

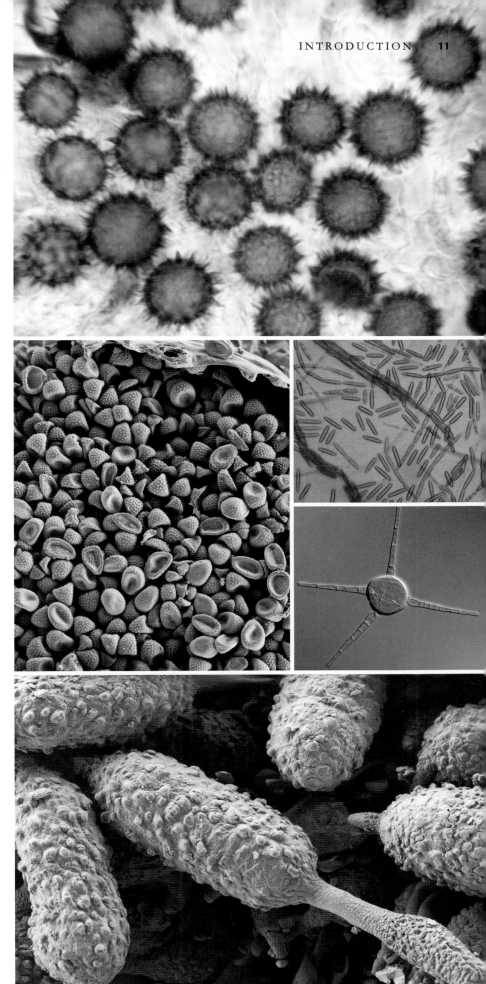

Right: Fungi are spread long distances by spores, equivalent to the seeds of flowering plants. They show huge variation in size and shape, and surface ornamentation.

Clockwise from top: Spiny spores of *Scleroderma areolatum*. Image: © Malcolm Storey, www.bioimages.org.uk Sausage-shaped spores of a *Cordyceps* species. Image: © Patrick Hickey. Star-shaped asexual spore of the aquatic fungus *Brachiosphaera tropicalis*. Image: Reproduced with kind permission of Elsevier. Spores of *Puccinia* species; Urediniospores of a rust fungus *Puccinia* species erupting through the leaf surface. Images: © RBGE/Stephan Helfer.

Fungi achieve this by secreting appropriate enzymes into their surrounding environment. Fungal feeding is therefore sometimes described as 'external digestion', which contrasts with the internal digestion of higher animals. Fungi secrete a vast range of enzymes and these allow digestion of the multitude of organic resources available on planet Earth.

The fourth defining characteristic of fungi is that they produce a huge range of different types of spores and these are involved in asexual or sexual reproduction. Although some lower plants, such as algae, mosses and ferns, form spores, they do not produce the extraordinary array of different spore types with the multitude of functions that are characteristic of fungi. Besides asexual and sexual reproduction, fungal spores are important for dispersal and survival.

Fungal diversity

Until recently the fungi were grouped according to the types of spores they produced. There were four main groups, with the rather unattractive names basidiomycetes, ascomycetes, zygomycetes and mastigomycetes. It soon became clear that the zygomycetes and mastigomycetes were a 'hotchpotch' of fungi that looked vaguely similar at one stage of their lives, but actually each contained several groups of fungi having very different structural characteristics and behaviour. Also, there was a fifth group called the deuteromycetes (or imperfect fungi) that was a 'ragbag' of fungi that never or rarely have sex. Now, however, following recent comparisons of the DNA sequences of hundreds of different fungi, we know much more clearly which fungi are closely related and which are not, and have a much greater understanding of their evolutionary relationships. There is now no need to have a group of 'imperfect fungi' as it has become clear that they virtually all belong to the basidiomycetes and ascomycetes.

The new groupings, with equally unwieldy names, still include the old favourites *Basidiomycota* and *Ascomycota*. The old zygomycetes have been split into the *Zygomycota*, which includes the pin mould on bread, and the *Glomeromycota*, containing fungi that have intimate associations with plant roots aiding water and nutrient uptake from soil (see Chapter 3). The mastigomycetes have also been split up; one group – the *Oomycota* – is no longer classified in the Kingdom Fungi as these fungus-like organisms are much more similar to algae. They are nonetheless very important organisms, as many are significant crop pathogens (see Chapter 3). Members of the new *Chytridiomycota* group, from the original mastigomycetes, are unusual amongst the fungi in having spores that swim using whip-like flagella at one stage in their life cycle.

DNA analysis has shown that a group of unicellular, parasitic microorganisms called *Microsporidia*, originally thought to be protozoa, are actually fungi.

Finally, a weird group of fungus-like organisms – the slime moulds – that were originally classified as fungi turn out not to be fungi at all but share features in common with fungi, animals and plants. Furthermore, they have different evolutionary origins and have now been placed in more than one group outside the Kingdom Fungi. The vast and intriguing diversity of fungi is explored in Chapter 1.

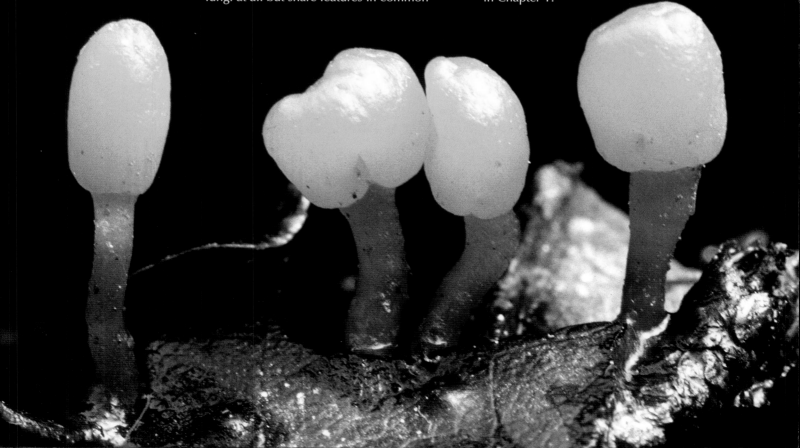

How fungi feed

Many fungi feed as microscopic yeast cells. Yeasts are very common in the natural environment, especially in sugar-rich situations such as on the skin of grapes and apples, and some such as *Candida* are also commonly found in the moist orifices of our bodies. Most yeast cells feed on simple sugars which they absorb. After they have reached a certain size they reproduce, most commonly by budding but sometimes by binary fission, in which the cell splits in two. When yeasts grow in the presence of oxygen they produce carbon dioxide, which is what makes bread rise. Other by-products of yeast growth can contribute to the flavour of bread. Yeast cells grown in the absence of oxygen produce alcohol (ethanol) as a result of fermentation. Wild yeasts found naturally on the skin of grapes can perform fermentation, but many commercial alcoholic products are produced using standard strains, such as *Saccharomyces carlsbergensis* used in lager production. Yeasts are probably among the earliest domesticated organisms, having been used by the Ancient Egyptians in bread making.

The majority of fungi are filamentous, forming hyphae and thus mycelia. The first stage of mycelium formation involves spore germination and the emergence of a filamentous, tip-growing hypha. Digested food is taken up as the hypha grows and branches. The hyphae and their branches tend to avoid

Left: The surface of fruit, such as the skin of grapes, is often rich in sugars on which yeast cells can feed. Image: © RBGE/Lynsey Wilson.

each other and thus space themselves out at the growing front of the mycelium. This feature allows hyphae to search the environment efficiently for food whilst reducing the problem of adjacent hyphae competing for the same food. Feeding by absorbing nutrients from the surrounding environment requires a large surface area to volume ratio, and long thin hyphae are therefore ideal. Growing as hyphae, rather than single cells, also means that fungi can penetrate into solid bulky resources rather

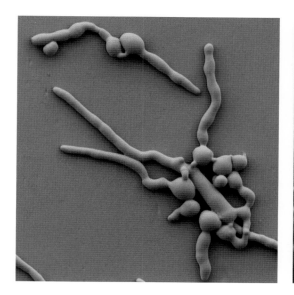

Far left: Germinating asexual spores of *Neurospora crassa*. Young hyphae grow away from the spores to establish the young fungal colony whilst other very short hyphae fuse to produce an interconnected network of germinated spores. Image: © Gabriela Roca, Chris Jeffree and Nick Read.

Left: The mycelial cords of two fungi showing interactions on contact. One species is growing from the left and can be seen ramifying over and under the cords of the other species in its search for food. At the points of contact droplets containing antagonistic chemicals have been exuded. Image: © Tim Rotheray.

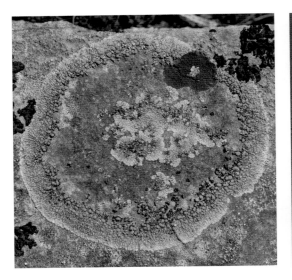

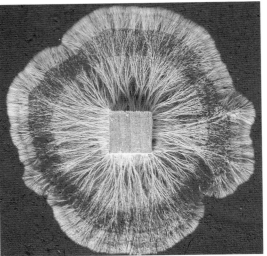

than just remaining on the surface. This is particularly important for those fungi that invade dead and living tissues of animals and plants in order to obtain nutrients. It is the mycelial way of life that sets fungi apart from unicellular microorganisms and makes them so important in terrestrial environments.

Growing outwards from a single origin (the spore) results in typically circular colonies. The youngest hyphae are on the outer margin, and the oldest in the centre. Behind the growing mycelial front some hyphal branches fuse with each other to form an interconnected network of hyphae. This feature greatly facilitates the transport of nutrients and water throughout the mycelium. Older hyphae in central regions eventually die and their contents are transported through the hyphal network

back to the growing front and are used in the production of new hyphae. The typically circular mycelia are visible not only on the agar jelly or trays of soil used to culture fungi in the laboratory, but also in nature. One example is the toadstools forming 'fairy rings' at the periphery of circular subterranean colonies (see Chapter 6). Another example is the circular patches produced by ringworm infections on human skin (see also Chapter 4).

Besides existing as clearly discrete filaments, hyphae under some circumstances converge and aggregate to form complex reproductive fruit bodies such as toadstools. In some species hyphae also adhere to form long linear organs called cords, or thicker, pigmented rhizomorphs resembling bootlaces. These structures are made up of bundles of hyphae arranged more or less in parallel. They are commonly produced by those fungi that rot wood and those called mycorrhizal fungi that help feed the roots of forest trees (see Chapter 3). Cords and rhizomorphs allow fungi to search for food, and to transport water and nutrients over long distances. Cords operate like a net – when a branch falls to the woodland floor it lands on the network, and fine hyphae grow from the cords and into the wood, incorporating the branch food resource into the system of interlinked cords. Mycelial networks provide the fungus with a large number of alternative routes to send nutrients from A to B. This is very important if part of the network is damaged by, for example, grazing soil invertebrates or even mammals (see Chapters 1 and 4).

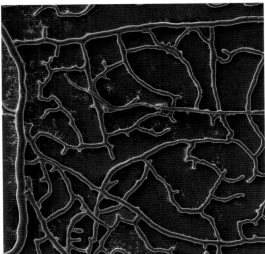

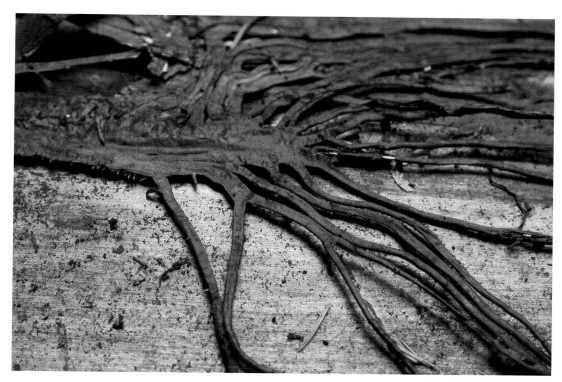

Left: The mycelium of some fungi grows as root-like organs (rhizomorphs) that can form extensive systems on the forest floor. Systems of some honey fungi (*Armillaria* species) probably form the largest organisms on the planet. Many honey fungi are necrotrophs – fungi that kill cells and feed off the dead remains. Image: © Patrick Hickey.

The extent of the network created by a fungal mycelium can be extraordinary. In popular quizzes the question 'Which organism is the largest on the planet?' is sometimes asked. A common answer is 'The blue whale'. The probable true answer is actually a fungus. In North America there are several examples of honey fungi (*Armillaria* species) whose individual rhizomorph systems occupy many square kilometres beneath the litter layer on the forest floor. If the whole rhizomorph system from a single fungal individual were weighed it would be considerably in excess of the weight of a blue whale! Furthermore, these humongous fungi rival trees in terms of what is the oldest organism on the planet, as some of these fungal networks are many hundreds, perhaps even thousands, of years old.

Overall, fungi have three main ways of obtaining nutrition: (1) by feeding off the dead remains – saprotrophy – of plants, animals and microbes, including other fungi (see Chapter 2); (2) from living cells – biotrophy (see Chapter 3); and (3) by killing cells and

Below left: *Erysiphe pisi* is a powdery mildew that is parasitic on members of the pea family. Powdery mildews are biotrophs – fungi that obtain food from living cells. Image: © RBGE/Debbie White.

Below: *Collybia confluens* is an example of the many fungi that are saprotrophs – fungi that feed on the already dead remains of other organisms. Image: © Chris Jeffree.

Fungi spread long distances as spores. Though spores are microscopic they can be seen when released en masse. Here a cloud of spores is released from a Puffball (*Lycoperdon perlatum*). Image: © Robert Pickett/ CORBIS.

Right: Spores can also be seen when a mushroom fruit body (with its stalk removed) is placed on a piece of paper away from draughts. After a few hours it will drop its spores revealing the pattern of the gills from which the spores were deposited. Image: © Patrick Hickey.

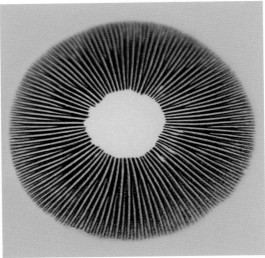

tissues and feeding on their dead remains – necrotrophy (see Chapters 3 and 4). Biotrophs damage hosts when they are parasites, but others provide benefits and are mutualists. These latter include lichens, which are organisms comprising a fungal partner in close association with an algal or cyanobacterial (blue-green algal) partner, and mycorrhizas, which involve intimate associations between plant roots and fungi (see Chapter 3). These three main types of nutrition are not mutually exclusive: for example, many mycorrhizal fungi have some saprotrophic ability, some mycorrhizal fungi can become pathogenic, and necrotrophs continue to live, for varying times, following death of the host.

How fungi spread

Fungi obviously spread as they grow through and between their food sources as hyphae, cords and rhizomorphs. They can also spread long distances as spores. These spores are microscopic, but we can see them when they are discharged *en masse* or fall together to form a spore print. Some types of spores are produced as part of sexual reproduction and others without sex. The large fruit bodies with which we are familiar – mushrooms, toadstools and brackets – are all produced as part of the sexual reproduction process (see Chapter 1). It is within these that the sexual spores are produced, shot away from the gills or pores and then spread on the wind.

Wind dispersal is only part of the story. Some spores are spread by hitching a ride on insect bodies; others are spread by slugs and snails and will only germinate to produce hyphae after passage through the animal's gut. Yet other fungi that have evolved an intimate association with insects (see Chapter 4) are carried in special pouches on the insect's body. Fungi which colonise organic matter in lakes, rivers and seas have evolved to be dispersed by water – they often have long projections that keep them suspended in the water and provide a stable fixture when they land on a suitable food resource. Though many spores are minute with thin walls so that they are easily spread by wind, some have very thick walls and are adapted to survive for many months, for example through adverse climatic conditions and shortage of food.

Production and destruction

Though we often only notice fungi when they rot our fruit and vegetables and other stored food products, they are an important food source for many animals, including man. Animals eat fungi because they are highly nutritious. They are high in protein and low in fat. All over the world people eat fungi which grow in the wild (see Chapter 9) or are cultivated (see Chapter 8). The mushrooms traditionally cultivated in Europe (*Agaricus bisporus*) account for only about 40% of

mushroom sales worldwide. The wood-rotting oyster fungi (*Pleurotus* species) are now increasing in popularity, as are the paddy straw mushroom (*Volvariella volvacea*) and shiitake (*Lentinula edodes*) that are favoured in Asian cuisines. Meat substitutes are made not only from soya but also from fungi – Quorn® is a filamentous fungus (*Fusarium venenatum*) formed into the texture of meat.

Yeast (*Saccharomyces cerevisiae*) has been used to brew beer and wine, and in bread making, for thousands of years. Soy sauce is another fermented fungal food. The flavour and fragrance of 'blue cheese' is produced by *Penicillium roqueforti*, and the surface rind of Camembert is also a result of fungal growth. In fact, nowadays a fungal peptidase is used

Above: The chanterelle (*Cantharellus cibarius*) is one of the more highly prized edible mushrooms harvested from the wild. Fungal fruit bodies, either cultivated or wild collected, are eaten by humans in large quantities around the world.
Image: © Chris Jeffree.

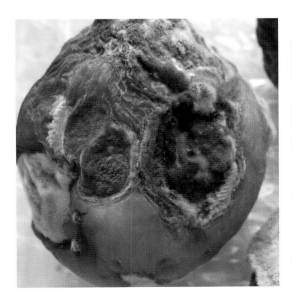

Far left: Fungi cause serious economic damage by rotting our food. This peach has been colonised by prolifically sporing ascomycete species. Image: © Caroline Hobart.

Left: Colonisation by fungi has been harnessed as a means of food production. Blue cheeses gain their flavour from *Penicillium roqueforti*. Useful compounds manufactured by fungi are also widely employed in the food industry. For example, almost all cheese is now produced using the fungal enzyme chimosin instead of rennet. Image: © RBGE/Lynsey Wilson.

instead of rennet to coagulate the curds in the earliest stages of production of most cheeses. Fungi that grow on the outside of cocoa beans produce the characteristic flavour of chocolate! Fungi also make the acidity regulator citric acid, used in soft drinks.

Drugs and diseases

Some fungi grow on us! Some are very irritating, such as those breathed in causing allergic reactions like hay fever and asthma, and those that grow on our bodies causing ringworm and athlete's foot. The latter two are caused by fungi that feed on the protein keratin found in skin, hair and nails. At the height of the mining and steel industries more 'sick days' were attributed to athlete's foot than to the common cold. Another fungus (*Candida albicans*) causes thrush, and yet another causes dandruff. A few fungi cause more serious problems, invading tissues inside our bodies (see Chapter 4). Fungal diseases are particularly prevalent in immuno-compromised patients, and mortality from human fungal diseases is increasing, mainly due to AIDS and the use of immunosuppressive drugs (e.g. to treat cancers and for stem cell and organ transplant operations). There are few drugs to treat systemic fungal human infections, primarily because the biochemical machinery of fungi and humans is very similar.

Fungi are amazing chemists. In rare cases they produce deadly poisons, such as the aptly

named death cap (*Amanita phalloides*). More typically, poisonous substances produced by fungi – mycotoxins – are not immediately fatal, but still of serious concern. The aflatoxins produced by *Aspergillus flavus* growing on stored peanuts and grain are a significant cause of liver cancer in parts of Africa. These negative impacts, as far as we humans are concerned, must be set against the fact that fungi have produced several of the 'wonder drugs' of the age (see Chapter 5). Penicillin, produced by *Penicillium chrysogenum*, was discovered by Alexander Fleming and saved many lives during World War II and subsequently. Statins, used to reduce our cholesterol levels, are also a fungal product and are the most widely used pharmaceutical drugs in the Western world – a £25 billion per year industry. In addition, fungi are important in the production of cyclosporins, which are used to prevent tissue rejection in transplant operations.

Useful killers?

All insect species suffer from fungal diseases, and fungi can be used as biocontrol agents of insect pests (see Chapter 4). Formulations of certain fungal pathogens of insects (e.g. aphids and locusts) are now commercially available, and are more environmentally friendly than chemical pesticides.

On the downside, many amphibian species are becoming extinct and many others are threatened with extinction due to infection by a chytrid fungus with the virtually unpronounceable name *Batrachochytrium dendrobatidis*, or *Bd* for short (see Chapter 4). Alarmingly, the first cases of this disease in the UK have recently been reported. No other

disease is known to have resulted in the mass extinction of a group of organisms on this scale. The problem seems to stem partly from global climate change providing conditions that allow the fungus to flourish and infect amphibians, and from the international trade in amphibians. The *Bd* problem is raising alarm bells with regard to the effects of climate change on biodiversity due to its influence on the distribution and activity of disease-causing organisms such as fungi. Many fungi too are endangered by environmental change, and if anything they are in greater need of protection than animals because of the vital roles they play in the functioning of ecosystems (see Chapter 10).

A bunch of rotters

Although some fungi destroy the fabric of our homes (e.g. the dry rot fungus, *Serpula lacrymans*), fungi that feed on dead tissues are the main garbage disposal agents of the natural world (see Chapter 2). They are the ultimate recyclers. When plants and animals die their dead tissues must be broken down to release the carbon and nutrients locked up in them, otherwise ecosystems would run out of nutrients, and plants and animals would die out. As part of the breakdown process, fungi form humus and improve soil quality, and they are major operatives in the compost heaps in

Above: Some fungi have a direct impact on humans by breaking down timbers in buildings and other man-made structures. The dry rot fungus (*Serpula lacrymans*) is particularly hard to treat. One of the tell-tale signs of this fungus is large fruit bodies with vast quantities of rust coloured spores. Image: © Nia White.

Left: Without the recycling activities of fungi the world's ecosystems would soon run out of nutrients and plants would not be able to grow. This mass of fruit bodies of a *Psathyrella* species is growing on woodchip. Image: © Patrick Hickey.

Elms infected with the Dutch elm disease fungus are ring-barked in an effort to prevent the disease spreading from the tree canopy, where infection begins, into the root system and potentially to other neighbouring trees. Image: © RBGE/Max Coleman.

Right: Some species of fungi interact with green algae or cyanobacteria to form a completely new organism – a lichen. Shrubby lichens, such as *Usnea* species, hang from the trees in the clean air and moist conditions of the west coast of Scotland. Image: © RBGE/Vlasta Jamnický.

our gardens. Because some fungi have enzymes that can break down complex chemicals, they have great potential for bioremediation of sites polluted with insecticides, herbicides, toxic metals, crude oil and creosote. They may even be used to remove radioactive waste (e.g. depleted uranium in battlefields) in the future. Fungi can break down some inorganic compounds too – for example stone, concrete and glass, and thus buildings and monuments.

Give and take

There is no doubt that some fungi can be a considerable nuisance to man, causing around 70% of crop diseases worth hundreds of billions of pounds worldwide every year (see Chapter 3). They have changed our landscape; for example, Dutch elm disease (caused by *Ophiostoma novo-ulmi*) has destroyed our mature elms, reducing this tree to a hedgerow species. They have even changed the course of history: late potato blight, caused by the fungus-like *Phytophthora infestans*, resulted in the Irish potato famine in the 1840s, leading to the death of nearly a quarter of the population due to starvation, and emigration of over a million people, mostly to North America.

Left: Turning the massive compost heap at the Royal Botanic Garden Edinburgh. During composting considerable heat is generated and on cool days compost heaps can visibly steam. Fungi are important decomposer organisms in compost. Image: © RBGE/Tony Garn.

However, it was fungi that allowed plants to colonise land about 450 million years ago by developing intimate partnerships with them and allowing them to access essential nutrients (e.g. nitrogen and phosphorus) they could not readily obtain by themselves from the terrestrial environment. Nowadays, about 90% of plant species in nature are fed water and mineral nutrients by mycorrhizal fungi, and without them would not survive or would have great difficulty in competing with other plants. The fungi benefit from this mutualistic association through the plant providing them with sugars (see Chapter 3). The soil beneath a few square metres of forest floor contains sufficient length of hyphae to stretch around the Earth's equator!

As mentioned above, fungi also form close partnerships with algae or cyanobacteria to form completely new organisms – lichens (see Chapter 3). The fungus provides the algae/cyanobacteria with minerals and physical protection; the algae/cyanobacteria provide sugars and sometimes nitrogen. Lichens are slow-growing, long-lived and often sensitive to air pollution, making them useful indicators of environmental change. Lichens can exist in very hot, cold and dry environments, are the main flora of the Arctic tundra, and have been shown to be able to survive outer space! They often occupy nutrient-poor environments (e.g. on rock), and are often pioneer species on bare rock.

The world of fungi is truly amazing, with new discoveries continually being made. What we have come to appreciate from the knowledge amassed so far is that without fungi our lives would not be the same. It is probably fair to say that we could not exist without fungi. Our land-based ecosystems simply would not work. We must safeguard fungi both for their future as well as ours (see Chapter 10). Our perception of fungi is complex and goes back millennia (see Chapter 6). In the science fiction literature of the 19th and 20th centuries, fungi have often been portrayed as a dark, malevolent force (see Chapter 7). It is true that fungi can be both friend and foe, but perhaps we have taken the negative view of fungi too much to heart. At a time when 'sustainability' is a modern mantra we would be wise to remember that fungi underpin the nutrient cycling that sustains life on Earth.

Main photos, clockwise from this page:

The slime mould *Stemonitis fusca* fruiting on bark. Myxomycetes are not actually fungi, but the fruit bodies look superficially like fungi.
Image: © Penny Cullington.

Orange-peel fungus (*Aleuria aurantia*).
Image: © RBGE/Robert Unwin.

Bird's-nest fungus (*Cyathus striatus*).
Image: © Chris Jeffree.

Facing page inset photos:

Velvet shank (*Flammulina velutipes*) (left);

Witches' butter (*Exidia glandulosa*) (right).

Images: © Peter Clarke.

Chapter 1
The Hidden Kingdom

Roy Watling

The Hidden Kingdom

Roy Watling

Above: In this illustration size reflects the diversity, in terms of numbers of species, within a particular group of organisms. The fungi, for instance, dwarf the mammals represented by the elephant. Speciescape: © Frances Fawcett.

U ntil fairly recently the fungi were considered a primitive, even 'Cinderella', group of plants. The boundaries of the group did not seem to be entirely clear, with some seeming to possess characters of simple animals, such as amoebae. We now know that the fungi are a huge group of diverse organisms brought together by their common method of obtaining food (see the Introduction) and quite distinct from both plants and animals.

Despite the past uncertainties about where fungi fit in the classification of life we have nevertheless relied on them over millennia for many familiar foods, drinks and chemical products (see Chapter 5). Fungi also underpin the process of nutrient cycling through rotting down plant and animal remains and making simple, easily used compounds available to other organisms (see Chapter 2). About 90% of vascular plants are tied to associations with fungi, the latter assisting in scavenging essential nutrients, helping their host plants to overcome competitors, and even protecting them from disease organisms (see Chapter 3). In spite of the ubiquity and importance of fungi, however, they remain little known to many people, with their beneficial activities being poorly appreciated. But perhaps this is not surprising, considering they are a largely invisible group of organisms.

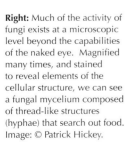

Right: Much of the activity of fungi exists at a microscopic level beyond the capabilities of the naked eye. Magnified many times, and stained to reveal elements of the cellular structure, we can see a fungal mycelium composed of thread-like structures (hyphae) that search out food. Image: © Patrick Hickey.

Far right: A nematode worm a millimetre or two long on a fungal spore mass. Image: © Chris Jeffree.

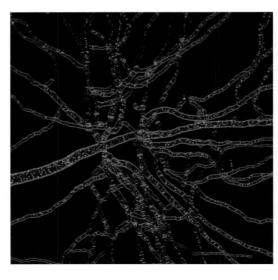

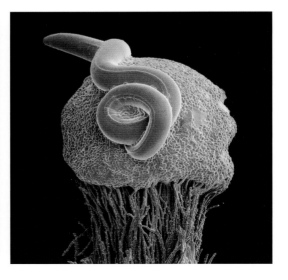

The reality is that if fungi stopped their work in nature things would be very different, and our broad canvas of human cultures would be unrecognisable.

There are estimated to be around 1.5 million species of fungi in the world. This enormous diversity makes fungi one of the largest groups of organisms known, dwarfing our own group, the mammals, which numbers a mere 4,260 species. Recent estimates suggest there are about 4,000 species of fungi associated with, or growing on, mosses alone. Similar numbers have been recorded from a handful of palm species. We probably know all but a few of the mammals, but perhaps less than 10% of the fungi. This discrepancy in knowledge is easily explained: mammals, in addition to being less numerous, tend to be large and conspicuous; fungi, in contrast, are often hidden from view for all or part of their lives. So where does this vast kingdom of life live, and what forms do its members take?

Hide and seek

A major reason why fungi remain largely hidden is that by far the majority are microscopic. We only become aware of their presence when our food goes mouldy, or we stop to think of the countless everyday products that require fungi in some way. Even those fungi which do produce visible structures, such as mushrooms, toadstools and brackets, remain hidden from view for all or much of their lives. They are essentially oversized 'microorganisms', as their life cycles are the same as the microscopic species. The visible parts are usually the fruit bodies produced for reproductive purposes. In a few cases these can be of giant proportions; one of the largest known weighs in at 45.5 kg and the biggest has a diameter of 4.9 metres. However, these are really exceptions to the rule, as in nearly all major ecosystems most fungi are to be found well out of human sight.

Above: *Clathrus ruber*.

Below left: Orange-peel fungus (*Aleuria aurantia*).

Below right: Bird's-nest fungus (*Cyathus striatus*).

Images: © Chris Jeffree.

The fungi show enormous diversity of form, including beautiful complex structures that sometimes display striking colours.

Above: *Russula atropurpurea*. About 90% of plants have fungi associated with their roots forming a partnership of mutual benefit, termed a mycorrhiza. The fungi provide the plant with water and mineral nutrients, while the plant supplies the fungus with sugars. Image: © Chris Jeffree.

Fungi occur in just about any environment you care to think of. They are to be found, for example, in the guts of insects, the stomachs of herbivores, inside tree buds and on the surface of roots, to name but a few. Anything from cracks in rocks to the few areas of open water in the Antarctic can provide suitable conditions; there are even aquatic fungi! Fungi are found literally everywhere and are especially important in forming close relationships with other organisms such as plants and insects (see Chapters 3 and 4).

Do flying fungi sound improbable? There are in fact species of fungi that manage to fly by attaching to the legs and bodies of insects, and in special pouches in some beetles. In the latter case, the fungi then colonise the beetles' wood-burrows, and the beetle larvae feed on the fungus as it is able to obtain nutrients from the wood that the insect cannot get on its own. Similarly, various tropical termites and ants cultivate specific fungi in special gardens in their nests, utilising the fungus to break down the vegetable matter they collect (see Chapter 4). The fungus is then used to feed the developing larvae. (And humans thought they were the first to domesticate organisms!)

Many fungi are hidden from view within the soil. Fungi are an important part of the soil ecosystem, providing food and shelter to soil-dwelling invertebrates. Some even turn the tables on animals and produce snares and sticky traps to catch soil microbes (see Chapter 4), something that probably had an origin some 100 million years ago.

Some fungi are parasitic. Parasitic fungi attack a wide range of organisms (see Chapters 3 and 4). On vascular plants they can produce spots or splodges on leaves and stems, and can even form red or yellow flower-like tongues on fruits and cones. White downy or powdery surfaces usually indicate the presence of their fruiting stages, while cereal heads and some flower parts may be reduced to powdery masses of spores (colloquially called bunts or smuts as they discolour the fingers when handled). More dramatic are the red, raw rings on human skin caused by ringworm (*Trichophyton mentagrophytes*). Other parasitic fungi have been implicated in the demise of many of the world's amphibians and even immunosuppressed hospital patients (see Chapter 4).

Right: Tar spot (*Rhytisma acerinum*) is very common throughout the range of the sycamore. As it is susceptible to pollution it is an indicator of clean air. Image: © Chris Jeffree.

Far right: Fungi can cause considerable food crop losses. Ergot (*Claviceps purpurea*) is a parasite of grasses and some cereal crops. Image: © Patrick Hickey.

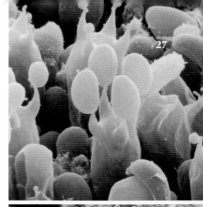

Left: Fungi can decay stored timber and timber in buildings. *Ganoderma colossus* growing from seasoning rubber timber in Malaysia. Image: © Roy Watling.

Right: The spore-producing surface of basidiomycete fungi can take many forms including gills, pores and various types of folds, but all are united in producing their spores in groups of four on the projections of a specialised cell called a basidium. From top to bottom: Basidiospores. Image: © Ingo Nuss. *Sparassis crispa*; *Datronia mollis*; *Boletus luridiformis*; *Lactarius* species showing a white 'milky' fluid that is produced following damage in these fungi. Images: © Chris Jeffree.

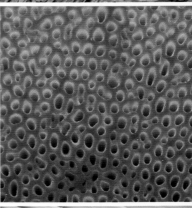

Fungi will also rapidly exploit new opportunities we create for them. The widespread use of wood mulch on flower borders, a popular activity in Britain these days, has attracted a whole new range of fungi to our gardens. The use of imported timbers in domestic dwellings has also brought its own problems; some of the fungi found destroying wood originate from abroad, while others have migrated from their usual habitats in woodlands to take up new homes in our houses.

Our breathing space

Plants produced the early oxygen-rich atmosphere through photosynthesis and paved the way for the evolution of larger, more energy-demanding organisms. None of this would have been possible without the fungi which developed intimate, mutually beneficial associations with plant roots. These associations are termed mycorrhizas (see Chapter 3) and they fall into two main groups. The first group – arbuscular mycorrhizas – consists of microscopic species found within the cells of roots and connected to the surrounding soil by thin threads through which they scavenge for nutrients. Many of our staple foods – including wheat, maize, millet, barley, oats and rice, and soft and stone fruits – have such associations. The other large group (numbering some 25,000 species, most of which are mushrooms and toadstools) is the ectomycorrhizas. These fungi form a sheath around the root of the tree, with some fungal filaments venturing between but not within the cells. Again, important nutrients flow between the partners, and the success of our dominant trees, both temperate and tropical, depends on such associations.

One of the most successful fungal associations worldwide – found from sea level (even in the sea) to the highest peaks and from the cold deserts to the tropics – is the lichens (see Chapter 3). Once thought to be independent organisms in their own right they are in fact an intimate relationship between one or more algae (or cyanobacteria) and a fungus. Lichens form the familiar colourful blotches on rocks, gravestones and tree trunks. Indeed, lichens may have been the primary source of oxygen in the earliest oxygen-rich atmosphere on Earth.

Who's related to whom?

Traditionally the fungi were divided into five major groups, based on the structure of their reproductive organs and whether sexual spores were formed or not. Following research over the last 25 years this approach has changed dramatically, resulting in a whole range of separately defined groupings (see the Introduction). However, two major groups defined in the old texts are still prominent in the classification and encompass the fungi likely to be encountered by the casual observer.

The first group, which includes the familiar mushrooms, toadstools and brackets, is characterised by the reproductive process coming to fruition in the production of four and sometimes more spores on the apex of projections arising from a single cell (a spore is equivalent to a seed in a flowering plant). This is exemplified by the fungi with an umbrella-shaped cap elevated on a stem, with the reproductive tissue beneath. In some species, perhaps more distantly related, this cell is divided, either transversely or longitudinally. A good example of the former is the jelly fungus Judas' ear (*Auricularia auricula-judae*) and of the latter the rust fungi which can devastate cereal

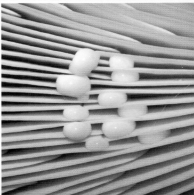

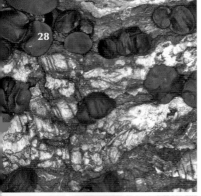

crops. All these fungi are called basidiomycetes after the single cell on which the spores are borne, which is termed a basidium.

The second group is sometimes referred to as the elf-cups and relatives, and these fungi have a distinctly different spore-bearing structure. The spores, generally numbering eight (though there may be multiples of this number), are produced in a sac (sometimes resembling a row of peas in a pod). In some species the sac is thick-walled or resembles a 'party popper', ejecting the spores with great force after the inner wall has become deformed by pressure. The group is divided further through different shapes and structures of the sac. The ascomycetes, as these fungi are known, derive their name from the spore-containing sac, or ascus. More than 30,000 scientifically described species have been catalogued, including the truffles, morels, powdery mildews, elf-cups and the Dutch elm disease fungus (*Ophiostoma novo-ulmi*).

Both these groups contain species which are known to have reduced their morphology to such an extent that they really only consist of egg-shaped cells. These are the yeasts, and they may be of ascomycete origin (with sacs) or basidiomycete origin (with the spores borne on a basidium). Thus even in a single group there can be enormous diversity of form. It is little wonder that the fungi have been so hard to understand. Many ascomycetes and basidiomycetes have lost the ability to produce their characteristic sexual spores, or do so rarely. Instead many different kinds of asexual spores are produced, varying in structure, colour and ornamentation. In the past the asexual fungi were classified together, but this grouping is now recognised to be unrepresentative of the evolutionary relationships.

Recent work has helped to clarify the relationships of several smaller groups that have very simplified external form and consist of little more than aggregations of

filaments or cells, which at most form swellings full of spores. Despite their simple form, their chemistry and cell components are very diverse because during their evolution they have exploited so many different habitats. They are generally microscopic, but the common pin mould (*Mucor*) is an example of one that produces a visible structure. Some microscopic species (chytrids) are motile at some stage of their lives, and have small filaments which act as propellers. Other superficially similar organisms (*Oomycota*) are now known not to be fungi and are considered to be related to brown algae and diatoms. They have lost the ability to make food using sunlight and now have a fungus-like lifestyle.

Another group that were traditionally regarded as fungi, but are now known not to be, are the slime moulds. Reports of white or yellow mucus climbing over vegetation represent sightings of the mobile stage of slime moulds. When mature they dry out and produce masses of powdery spores which cling to objects (especially the fingers!). They are also known to burst out of the spore on moist human skin and try to enter hair follicles, causing a very itchy sensation.

Left: Ascomycete fungi produce their spores in a pod-like structure called an ascus (plural asci). The fruit bodies that house these asci are, however, varied in form depending upon the species. From top to bottom: *Bulgaria inquinans*; purple jellydisc (*Ascocoryne sarcoides*); jelly babies (*Leotia lubrica*); candle-snuff fungus (*Xylaria hypoxylon*); eye-lash fungus (*Scutellinia scutellata*). Images: © Chris Jeffree.

Right: The ascus is a pod-like structure that contains the spores of ascomycete fungi. Typically there are eight spores, but sometimes multiples of eight are seen. Image: © Nick Read.

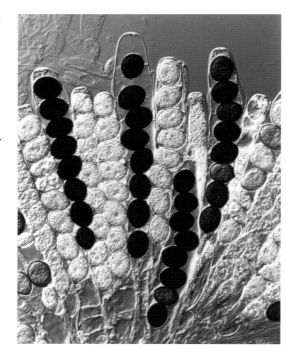

Left: Porcelain fungus (*Oudemansiella mucida*) releases white spores that can be seen covering the wood below the fruit bodies. Image: © Chris Jeffree.

Getting about

Fungi that produce obvious fruit bodies (mushrooms and toadstools) are primarily composed of filaments that grow in the soil or organic matter, spreading over an ever widening area. These filaments are termed hyphae and when bunched together they form what is called a mycelium (see the Introduction). The filaments are hard to see but the mycelia take on a distinct pattern – for example the dreaded and distinctive mycelia of the dry rot fungus (see Chapter 2) growing over woodwork, plaster or brickwork as fans. Mycelia may exist for many years rarely giving away their presence, and the fungus may only become obvious when favourable conditions result in sexual or asexual reproduction (though for many species this also often occurs out of sight in nooks and crannies). The culmination of reproduction is the production of microscopic spores, which are then dispersed by wind or rain-splash, physical disturbance or insect or larger animal vectors – in their stomachs, on their feet or by simply hitching a lift on their bodies. When the spores find a suitable resource on which to germinate, the process starts again. Some fungal spores, when present in great quantity, can cause asthmatic attacks in susceptible individuals.

Spores and hyphae can have a hard time in nature as they provide nutritious packets of food for microorganisms. Work in the Borders of Scotland, and elsewhere, is revealing the complex and previously unappreciated relationships between fungi and tiny soil invertebrates which exist beneath our feet. Hidden from view in the soil fungi are constantly scavenging for food. Even in the depths of winter when it is too cold for the soil animals to be active the fungi continue to grow. Fungal hyphae make an admirable meal for these animals including springtails, mites and nematode worms. As conditions warm these minute fungal 'grazers' begin to have an impact on the growth of fungi. The result is an unseen battle raging in the soil,

Left: Although spores are often transported by wind or water, the spores of some fungi are transported by animal vectors. For example, the Dutch elm disease fungus (*Ophiostoma novo-ulmi*) reaches new elms by hitching a ride on elm bark beetles. It infects healthy elms via feeding wounds that the beetles create on young twigs. Image: Forestry Commission Picture Library.

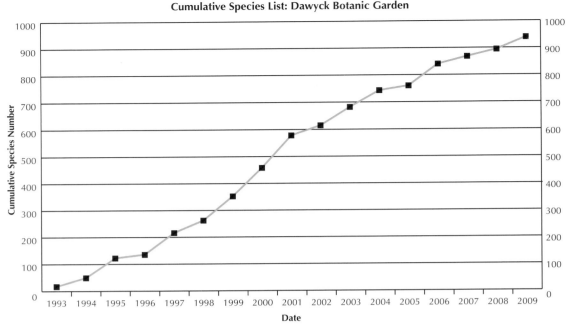

Cumulative Species List: Dawyck Botanic Garden

largely controlled by environmental conditions. Depending upon who has had the upper hand a wet autumn might produce an impressive fruiting of colourful mushrooms and toadstools, or simply a fungal 'whimper'. A dry summer favours those fungi that can transport water over considerable distances. Their grazers, on the other hand, are possibly limited by extended dry periods.

The sexual structures of the more obvious species may not be found every year, even in areas which are being closely monitored. For some species it is not unusual for the time between seeing a fungus in a particular location to be up to 20 years or more. This makes recording fungi a rather difficult and possibly scientifically hazardous occupation. However, help is at hand. The introduction of molecular techniques (studies of DNA and protein variation) has allowed the unravelling of evolutionary relationships between fungi (see the Introduction). A group of techniques called DNA fingerprinting make it possible to trace a fungus within wood or soil without having to have the tell-tale fruit body present. The individual molecular make-up of a fungal form can be checked against a library of molecular fingerprints, obtained from known species – like tracking down a criminal by their DNA. In this way we are able to understand more what the fungi are doing whilst out of view and trace where they are hidden. This technology has revealed a number of fungal record breakers. It is now known that some species of the honey fungus (*Armillaria*) can have individuals that cover about 900 hectares, and weigh over 600 metric tons. Some of these massive fungi are estimated to be around 2,400 years old! Such great age estimates contrast dramatically with the few days lifespan of the fruit bodies (mushrooms) produced by these fungi.

Ancient legacy

Fungi are an ancient group – the evidence suggests they may have been around for as long as 850 million years. The variation held in the genetic code (DNA) of living things can be used to estimate the time since two lineages diverged from a common ancestor. The assumption made is that the DNA being studied is accumulating changes (mutations) in a clock-like fashion. Using dated fossils or other datable events the rate of change can be established and estimates of time can be made.

Traces of what might be fungi can be recognised in rocks of most ages, even as far back as the sediments formed on the earliest shores some billion years ago. As the vascular plants evolved so did the fungi, and fungal remains can be found in many coal and peat-forming deposits the world over. A few have been preserved within the cells of primitive land plants; these not only resemble the form of those found today but also appear to have been playing the same role as present-day arbuscular mycorrhizal fungi (see Chapter 3). At about the same time, nearly 400 million years ago, there was what is sometimes considered to have been lichen growing up to 9 metres tall; it was never seen again in geological times. It has now been shown that it is more likely to be intertwined bundles of liverworts and fungal hyphae. Many of the fungal remains in rocks are in the form of spores, although impressions of woody fungi are known from as far back as 90–94 million years ago. The most spectacular finds have been in ambers 15–30 million years old. These show perfectly formed mushroom-like structures, demonstrating that this morphology has been around for a long time and at least since the dinosaurs roamed the Earth!

Hand-in-hand

Man has used a few larger fungi for specific purposes. Here it is not the fungus that is hidden but rather the knowledge of its use. Well-documented uses include horse's hoof fungus (*Fomes fomentarius*) for making belts, aprons and hats, and *Osmoporus* as a deodorant. The willow bracket (*Phellinus igniarius*) has been used to manufacture snuff, whilst puff- and fuzzballs (e.g. *Bovista nigrescens*) have been used to carry fire. Larch bracket (*Laricifomes officinalis*), among many others, was regarded as a medical panacea. American Indians have made use of fungi which possess fragrant odours

Left: Fossil bracket fungi from the Tertiary rock of Idaho in the western United States. Image: © Roy Watling.

Below: Horse's hoof or tinder box bracket (*Fomes fomentarius*) is widespread, especially on old birch trees. Image: © Patrick Hickey.

Hat made from
Fomes fomentarius
purchased by the author
in Szentendre, Hungary.
Image: RBGE/Lynsey Wilson.

Inset: Seller of items made
from *Fomes fomentarius*
in Szentendre, Hungary.
Image: © Roy Watling.

when dried, and several aboriginal peoples (including those in Australasia, the Americas and Africa) have used fungi to dye themselves.

Fungal remains found at archaeological digs can present a challenge. Although we may believe the fungus had a specific use there may be little evidence to confirm or refute this. Accompanying the Stone Age man Ötzi, found in melting ice at the edge of a glacier in the Alps, was a 'survival pack'. This contained a fungus probably used for tinder (tinder box bracket or horse's hoof, *Fomes fomentarius*), a fungus used for sharpening arrows and knives and as an antibiotic (birch bracket, *Piptoporus betulinus*) and another poorly preserved which it is speculated was possibly used as a talisman.

The hidden nature of fungi, combined with their rapid and unpredictable appearance, has led to many cultural associations which appear in literature and folklore (see Chapters 6 and 7). In certain European cultures these attributes have led to fungi being viewed negatively and associated with toads and reptiles and the underworld. Similarly, fungi have been linked with bats and nightjars, which fly at night. These animals are the natural analogue of the witches and warlocks of folklore and this association draws fungi into the realm of black magic. In addition, there were many unexplained phenomena which concerned early people such as rings appearing in grassland as if the result of wee people dancing in circles (see Chapter 6). What better links than all these do you require to substantiate a connection with the underworld?! So although most fungi are hidden, those which are more obvious have been a component of folklore throughout time.

The hidden nature of most fungi means their vital importance is easy to overlook. This book introduces the weird and wonderful world of fungi from a range of different perspectives to highlight how fungi touch so many aspects of our lives. It is arguable that we might not even be here were it not for fungi. Could life exist without fungi? This is a difficult question to answer. What we do know is that life without fungi would not be life as we know it.

Bovista nigrescens (brown puffball) species profile

With excitement archaeologists at Hadrian's Wall uncovered several small, squashed but apparently previously more-or-less spherical objects. However, excitement turned to disbelief when leather experts concluded that these possible containers were not in their arena. But to whose camp should the excavators go? Mycologists are frequently at the end of the line of enquiry, often sent things in desperation. And so it was that the specimens landed on my table. Examination showed that they were in fact relatively well-preserved mature brown puffballs, a common grassland fungus in the British Isles.

Being mature the specimens could not have been collected as food as they were full of spores. As a food they would have been white and immature, lacking spores, and they would not have matured once picked. You don't pick puffballs and then sort out the mature from the immature unless there is a use; it would just waste valuable time! Unbeknown to the archaeologists they had stumbled across a Roman 'chemist's store'. The puffballs were probably collected to staunch bleeding, if parallels can be drawn from modern tribal communities, though other uses are known – in some parts of Africa smouldering specimens are used to smoke honey bee nests and so pacify the bees.

The solving of the Hadrian's Wall mystery opened up the flood gates and puffballs began to be received from other perplexed archaeologists and excavations from widely scattered areas of the UK. Specimens of mosaic puffball (*Lycoperdon utriforme*) came from Skara Brae in Orkney in quite considerable numbers. So other species of puffballs were being used; and Skara Brae is thousands of years earlier than the Roman invasion. Other samples came from, for example, York and Norfolk.

Some interest was also shown by researchers from the Blood Transfusion Service as puffball spores are warty and about the same size as blood-corpuscles so they can act as a nucleus for blood-clotting – and they show no allergic activity.

But the story does not end there – further fungal material subsequently came in from all sorts of archaeological sites, from vitrified forts to Stone Age cairns, some even labelled 'knee caps'….

Roy Watling

Image: ©RBGE/Ross Eudall.

Main photos, clockwise from this page:

Calocera viscosa;
Lepista flaccida.
Images: © Peter Clarke.
Marasmius rotula.
Image: © Chris Jeffree.

Facing page inset photos:

Mycena riparia (left).
Image: © Jacob Heilmann-Clausen.
Lanzia echinophila (right).
Image: © Martyn Ainsworth.

Chapter 2

Recycling the World

Lynne Boddy, Geoff Robson &
Naresh Magan

Recycling the World

Lynne Boddy, Geoff Robson & Naresh Magan

We all know that fungi are rotters! We often see them making our bread and cheese mouldy, growing on the surface of jam, or rotting our fruit and vegetables. Most of us have been unlucky enough to find them rotting the fabric of our homes and gardens, perhaps causing wet rot or dry rot of timbers, as black moulds on shower curtains or damp walls, or rotting wooden window frames, fence posts, garden sheds and furniture. Clothing and the paper and bindings of books (if slightly damp), photographic slides, and audio and video-cassette tapes are not immune either, and fungi can even grow on camera lenses and in fuel. Millions of pounds are spent every year on preservation and renewal of decaying timber products.

In light of all this destruction we could be forgiven for thinking that fungi are a nuisance! However, we certainly could not live without them. The fungi that feed on dead organic matter such as dead animals, plants and materials derived from them, such as wood, paper and cotton, are the 'garbage disposal agents of the natural world' – they are the world's best recyclers.

When plants and animals die their dead tissues must be broken down to release the nutrients locked up in them, otherwise ecosystems would run out of nutrients and plants would be unable to grow. While bacteria and soil invertebrates such as earthworms, woodlice, millipedes, mites and springtails play a role,

Below: A collection of fungus fruit bodies, many of which are produced by fungi that decompose dead organic matter. Image: © Chris Jeffree.

it is almost exclusively fungi which are able to deal with the complex lignocellulose that makes up bulky plant material. To put this in perspective, every year plants grow and add a total of 56.4 petagrams (56,400,000,000 metric tons) of carbon to Earth's ecosystems on land. To maintain the balance in these ecosystems a similar amount is decomposed, releasing carbon dioxide, water and mineral nutrients, and this is achieved largely by fungi. As part of the decomposition process humus is formed, which acts as a reservoir of nutrients and helps to promote good soil structure in which plants can grow. In our gardens we make use of this fungal decomposer activity in our compost heaps.

By virtue of the ability of some decomposer (also called saprotrophic) fungi to break down very complex molecules, fungi are likely to be the answer to our problem of disposing of long-lived rubbish such as plastics, and they can also

be useful in cleaning up polluted sites. Some fungi that rot wood do so only slowly and can enhance the value of wood for production of furniture and decorative items. Others that are aggressive against fungal pathogens can be used to control such pests. These topics are touched on below, after a consideration of decomposer fungi that cause difficulties for man.

Decay of our homes

To a fungus the wood in our homes is a source of nutrition just like the dead wood and other organic matter in the natural environment. If wood in our homes becomes damp then fungi will start to decay it. Perhaps most feared is the so-called 'dry rot fungus' – *Serpula lacrymans*. In North America the term 'dry rot' is also used for the damage caused by *Meruliporia incrassata*. Actually the term dry rot is a misnomer as, like all fungi, *S. lacrymans* needs a source of water. It becomes established in wood dampened by, for example, a leaky roof or downpipe. It is then able to spread into dry timber by transporting water through its mycelial cords (mycelial cords are aggregates of the fine threads, or hyphae, that form the body of the fungus; see Chapter 1). Droplets of water are exuded from the tips of the hyphae, and it is from this that the fungus got the second part of its name – *lacrymans*, meaning crying. The fungus is extremely invasive, being able to grow through plaster and brick, and can potentially spread from one house to another

Above left: *Pluteus* species are often found decomposing wood that is already well decayed. *Pluteus leoninus* produces striking yellow fruit bodies on well-decayed wood. Image: © Stuart Skeates.

Above: The beef steak fungus (*Fistulina hepatica*) is fairly specific to oak and chestnut. It very slowly rots the central wood of trunks, and in so doing turns the wood a beautiful brown colour, much prized for decorative veneer. Image: © Stuart Skeates.

Left: The dry rot fungus (*Serpula lacrymans*), naturally occurring in only a few places including the Himalayas, now infects timber in buildings in many regions of the world. It causes major timber decay problems and is very hard to eradicate. Here, fans of mycelium are seen spreading away from colonised wood in search of fresh wood. Image: © Nia White.

in a terrace. This makes it very difficult and expensive to eradicate: all colonised wood must be removed, including at some distance beyond where it is evident. Mycelium in plaster and brick must also be killed and, above all, it is essential that there is no source of water.

Serpula lacrymans and some other more common decayers of wood in buildings, such as the 'cellar fungus' *Coniophora puteana*, cause brown rot. As the name suggests, the wood becomes brown and crumbly, with cubical cracking. This is because cellulose and hemicelluloses are broken down but lignin is not. The wood loses its strength rapidly, which is a major problem when the wood forms a structural component of buildings! White rot, caused for example by turkey tail (*Trametes versicolor*), also occurs in buildings. The wood becomes bleached, and ultimately all of the wood, including lignin, is converted to carbon dioxide and water, leaving nothing remaining. However, strength loss is less than in brown-rotted wood at early stages, when comparison is made between wood having the same weight loss. All of these fungi, except *S. lacrymans*, are common in the natural environment. *Serpula lacrymans*, which originates from the Himalayas, has found just the right conditions for itself in our homes,

where the microclimate is fairly stable and other competitors are infrequent.

Soft rot occurs under very wet or fluctuating conditions, and in preservative treated timber, where white- and brown-rot fungi have been inhibited. It is typically caused by sac fungi (ascomycetes; see Chapter 1), whose activity characteristically causes superficial softening of wood under water, and deeper softening elsewhere. The wood often becomes darker and cross-cracked when dry. Amazingly, wooden huts and other wooden items left in Antarctica, following Robert Falcon Scott's ill-fated expedition to the South Pole, now show signs of soft rot, despite the intense cold. Destruction of wooden archaeological artefacts by soft rot is common (e.g. the wood in the ancient tomb of King Midas in Turkey).

Decay of our food

People often ask whether we eat to live or live to eat? Whichever it is, we often find that some other organism has got to our intended food first. Fungi are ubiquitous as spores in the air and these can be deposited on growing food crops in the field, causing diseases and losses of yield and nutritional quality (see Chapter 3). Most raw food commodities such as grain, fruit and vegetables are alive and respiring, and their surfaces are colonised by a wide range of yeasts and filamentous moulds, even before they reach our kitchens. Green or blue moulds (*Penicillium* species) on damaged citrus fruit, apples and pears are examples of fungi that spoil produce. Staple cereals such as wheat, barley, maize and rice are also prone to mould spoilage, insect damage and sometimes spontaneous heating, if they are poorly stored, causing losses in nutritional quality.

Many raw foods are subsequently processed into products such as bread and cakes, jams, yoghurts, cheeses and ready to eat meals, and these products can become contaminated by moulds during processing if hygiene standards are not maintained. For example, yeasts can grow in poorly stored yoghurts, bread mould (*Mucor*) can colonise bread products, and *Penicillium* and *Aspergillus* species can colonise bread and other cereal-based products.

Certain spoilage fungi, including species of *Aspergillus*, *Penicillium* and *Fusarium*, are also able to produce toxic chemicals, called mycotoxins. The most important ones produced naturally in a wide range of food products include the carcinogen aflatoxin, and the possibly carcinogenic ochratoxin,

patulin and trichothecenes. Aflatoxins can be present in maize, nuts and spices. Ochratoxin can contaminate some cereals, raisins, cocoa and red wine. Windfall apples can become damaged and *Penicillium* infection can lead to the contamination of apple juice with patulin. Because these mycotoxins are heat stable and difficult to destroy there are strict legislative limits on the amounts that are allowed in food commodities.

Why does mould spoilage occur? The growth of moulds in food products is determined by some key environmental and chemical factors. In particular, warm, moist conditions are conducive to fungal growth. Preservatives are, therefore, often incorporated into food products to try to extend shelf-life. In Europe today, the main approach to minimising mould spoilage and mycotoxin contamination in the food chain is management based on 'hazard analysis critical control point' (HACCP) systems. This approach originates from the NASA space program, and involves identifying and monitoring the critical control points in a chain to minimise risk. Thus, in most food chain systems the critical control points in the process – pre-harvest, post-harvest and during processing – are monitored to minimise the risk of microbial spoilage and maximise the shelf-life of a food.

Above: As decomposers fungi can cause a problem to man when they rot our food such as this bowl of fruit and vegetables. Image: Reproduced with kind permission of www.fungi4schools.org © David Moore.

Left: One way in which we preserve food is to add large quantities of sugar, which makes it difficult for most microorganisms to get enough water to grow. *Penicillium* species can, however, grow well on jam. Image: Reproduced with kind permission of www.fungi4schools.org © David Moore.

Right: In moist conditions
fungi can grow on paper,
material and other decorative
items causing discoloration
and decay, as seen here
on this old book.
Image: © RBGE/Lynsey Wilson.

Nuisances in our everyday lives

Although fungi play a vital role as decomposers in ecosystems, we tend to notice them only when they grow where we do not want them to. Apart from growing on surfaces and materials such as plastics in our homes, fungi can cause problems in unusual circumstances. Some fungi can degrade diesel, petrol and aviation fuels when even small amounts of water and pockets of air are introduced into fuel tanks. Species of *Trichoderma* and *Mortierella* are the most commonly recovered species and can cause blocked fuel lines, with potentially catastrophic consequences. Camera lenses, CDs and DVDs can be vulnerable to attack as fungi can utilise the coatings and secrete acids that can etch deeply into the surface. Cinematographic film when stored is also vulnerable to contamination, and fungal growth can cause the film to be so badly degraded that it cannot be saved.

Books and works of art are equally vulnerable, which is one of the reasons why particularly valuable items have to be stored in humidity and temperature controlled rooms. Fungal growth is a particular problem in some of the tombs in the Valley of the Kings, where considerable and irreversible damage has been caused.

Fungi grow commonly on surfaces in our homes, wherever there is moisture, and we often refer to them as mould. The most common fungi found indoors are species of *Alternaria*, *Aspergillus*, *Cladosporium* and *Penicillium*. They commonly spread via spores. Some can produce toxins that are potentially harmful to health, for example black mould (*Stachybotrys chartarum*), causing allergic symptoms and respiratory problems. Under normal circumstances it is just necessary to take common-sense routine measures to prevent mould growth in the home – keep humidity low and remove surface growth with appropriate cleaning products.

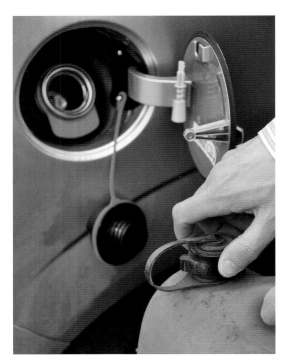

Right: Some fungi can
break down diesel, petrol
and aviation fuels, and
can cause blocked fuel
lines with potentially
catastrophic consequences!
Image: © RBGE/Vlasta Jamnický.

Recycling long-lived rubbish

Since the production of the first man-made plastic, Parkesine, in the 1850s (a material derived from cellulose that could melt upon heating and retain its shape after cooling), an array of chemically different plastics, largely derived from petrochemicals, have dominated and defined modern life. These possess a number of important attributes, including flexibility, durability, strength, lightness, corrosion resistance and low production costs. It is estimated that the worldwide production of plastic polymers is in excess of 260 million metric tons per annum.

Many plastics, such as polyvinylchloride (PVC, used in window frames and doors), polystyrene (used in packaging), polyethylenes (plastic bags) and polycarbonates (bottles), are highly resistant to any form of chemical or biological degradation, but additives mixed into these materials to change their physical properties are often vulnerable to fungal attack. For example, additives to make PVC flexible, enabling it to be used to make shower curtains, laminate floorings, bath mats and cable insulation, are vulnerable to fungal attack, while other plastics such as polyurethanes (often used to make foams for furniture, packaging and items such as shoe soles), polyamides, polyesters and polyvinyl acetates (often used in paints) can themselves be directly colonised and degraded by fungi. In bathrooms, the black yeast *Aureobasidium pullulans* frequently grows on the base of shower curtains and the underside of bath mats. Such is the vulnerability of these materials that biocides are incorporated during their manufacture to retard mould growth and increase the lifetime of the product.

The ability of fungi to attack and utilise many plastics and their additives reflects their success as saprotrophs, capable of secreting enzymes that have evolved to degrade a diverse range of natural polymers in the environment such as proteins, starch and lignocelluloses. As plastics have been in the environment for only a few decades, susceptible plastics and additives are the ones which contain chemical bonds that are similar to those occurring in nature. We can therefore look to the fungi to help us get rid of our man-made recalcitrant waste.

Left: Many plastics are highly resistant to chemical and biological degradation, but additives to change their physical properties can be directly colonised and degraded by fungi. Some plastics, such as polyurethanes, polyamides, polyesters and polyvinyl acetates, can be degraded by fungi. Fungi may, therefore, be able to help us get rid of our recalcitrant waste. Image: © RBGE/Max Coleman.

Mycoremediation

The ability of fungi to break down complex molecules might offer possibilities for 'cleaning up' some of the mess made by man. Some fungi can break down hydrocarbons (oil and petrochemicals). Others can enhance the degradation of herbicides and pesticides, including simazine, trifluralin and dieldrin. Uranium, from spent ammunition on battlefields, can be sequestered by fungi, and fungi are involved in recycling several metals (e.g. selenium and cadmium). However, although such mycoremediation holds promise, this area is currently still in its infancy.

Below: Some fungi can grow from one lump of rotting plant material to another, but others are confined to the food source that they are growing in and must spread as spores elsewhere before they 'eat themselves out of house and home'. *Lanzia echinophila*, an ascomycete cup fungus, fruits on old sweet chestnut husks and spreads to new husks exclusively by spores. Image: © Martyn Ainsworth.

The ear pick fungus (*Auriscalpium vulgare*) is a decomposer fungus that is often found on buried decaying pine cones. It is unable to search for new food resources by growing out as mycelium, but rather must spread as spores. Image: © Chris Jeffree.

Recycling dead bodies

It is said that we start to die from the minute we are born. In fact we and all other organisms, or at least dead parts, are decomposing continuously: dead epidermal cells are brushed off and hair and fur falls out; urine and faeces are deposited; plant roots exude chemicals and cells are sloughed off as they grow through soil; bud scales, fruits, leaves, twigs and branches fall. Ultimately, the whole body dies. While many different groups of organisms are involved in decomposition and nutrient release, fungi are often the most significant because of the huge breadth of enzymes produced, including those that break down complex molecules such as lignin, and the mycelial lifestyle that allows them to colonise bulky solid organic materials, such as fallen wood.

Different species of fungi are often found colonising different plant or animal species. Some are quite specific, such as the birch bracket (*Piptoporus betulinus*) on birch wood, whereas others are found on a wide variety of hosts, such as turkey tail (*Trametes versicolor*). Different species of fungi are often found in different parts of the animal or plant. Many fungi are confined within fallen plant parts, unable to grow out as mycelium and able to leave only when they produce spores in fruit bodies.

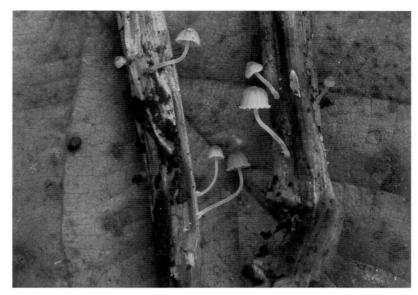

For example, various species of *Marasmius* are restricted to the stalks of leaves, *Mycena* species to the blades of leaves, *Mycena* species and the ascomycete *Xylaria carpophila* to beech cupules and the beef steak fungus (*Fistulina hepatica*) to heartwood. On the other hand, other fungi are not restricted to individual dead plant parts but can grow out and spread between different resources on the forest floor. For example, *Marasmius wynneae* colonises large patches of dead leaves and *Clitocybe nebularis* forms 'fairy rings' on the forest floor.

Above: *Mycena riparia,* a decomposer fungus that is associated with the leaves and bases of the sedge *Carex acutiformis* and closely related species. Image: © Jacob Heilmann-Clausen.

Left: *Marasmius rotula* decomposes small pieces of plant debris. Image: © Chris Jeffree.

Some fungi specialise in growing in and rotting wood at late stages of decay. The late stage basidiomycetes *Pluteus aurantiorugosus* and *Ossicaulis lignatilis* (rear) are seen here fruiting in a hollow elm log.
Image: © Martyn Ainsworth.

Right: Mycelial cords over the surface of fallen wood.
Image: © Jacob Heilmann-Clausen.

The fungi involved in decomposition of dead organic matter change over time, with certain groups of fungi predominating at early, middle or late stages. For example, in dead oak branches still attached to the trunk, the fungi that begin the decomposition process include *Stereum gausapatum*, *Peniophora quercina* and *Vuilleminia comedens*. They are fungi that are present as harmless, hidden endophytes – literally 'within plant' (see Chapter 3). These fungi existed within the sapwood when the branch was alive, and only grow as extensive mycelia when the branch dies and begins to dry out.

The early colonisers are then followed typically by turkey tail (*Trametes versicolor*), *Stereum hirsutum* or *Phlebia radiata* that arrive as airborne spores. They are antagonistic, being able to kill mycelium of less antagonistic fungi. Other less combative fungi are also found while branches are still in the canopy, such as *Hyphoderma setigerum* and *Schizopora paradoxa*, which though not good combatants are very tolerant of desiccating conditions.

When branches become sufficiently weakened by decay, they fall to the forest floor where the decomposition process continues. Other fungi may then arrive by growing through soil as mycelial cords. These are very combative and can replace all fungi that have arrived before them. These may be the final fungi in the succession, eventually completely decaying the wood. Sometimes fungi that specialise in coping with the dwindling

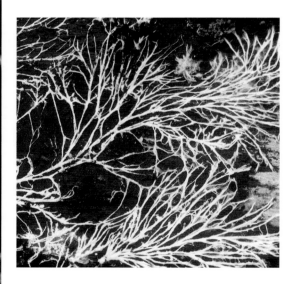

resources of well-decayed wood, such as *Pluteus* species, complete the succession.

Successions of fungi are found on all decomposing organic matter. Another example is decay of herbivore dung. This is brought about by invertebrates and fungi. There is a clear succession of fruit bodies: after 1–3 days species of *Mucor*, *Pilaria* and *Pilobolus* are present, but disappear after about 7 days; by 5–6 days species of the discomycetes *Ascobolus* and *Coprobia* appear, and are joined after 9–10 days by other ascomycetes, such as *Sordaria* and *Podospora* species; finally basidiomycetes like *Coprinus* species fruit. However, just because the fruit bodies appear in this order should not fool us into thinking that this is the order in which these fungi arrive or are most active. In fact, many of these species are present when the dung hits the ground (passing through the animal's gut as spores), or very soon after. The early fruiters are smaller and have less complex fruit bodies, and hence can acquire enough nutrients to produce fruit bodies more quickly than the basidiomycetes with their larger, more complex fruit bodies.

While dead organic matter made up of complex molecules such as lignin is rotted almost exclusively by fungi, the fleshy parts of vertebrates are usually decomposed by the body's own enzymes, by bacteria and by insect larvae. However, again, the more recalcitrant parts are typically dealt with by specialist fungi, for example *Onygena equina* on horn. Several species are found fruiting at early stages of

decomposition of the bodies of vertebrates, for example some *Ascobolus* species and the basidiomycete *Coprinus neolagopus*. Different basidiomycetes fruit at late stages of decomposition, including some *Hebeloma* species and *Laccaria bicolor*.

Above: Though soft parts of animal bodies are easily broken down by a wide range of microorganisms and invertebrates, only a relatively few microorganisms can make the enzymes necessary to break down horn. Here, the specialist decomposer *Onygena equina* is fruiting on horn. Image: © Penny Cullington.

Far left: Dung has its own communities of specialist rotting fungi. The microscopic asexual fruit bodies of the zygomycete fungus *Pilobolus* are shot up to a metre beyond the dung on which they grow, in order to escape the so-called 'ring of repugnance' around the dung. This increases the chances of the spores being eaten by grazing animals and so passed with the dung ready to germinate. Image: © Malcolm Storey, www.bioimages.org.uk

Left: *Coprinus sterquilinus* fruits after early stage decomposers like *Pilobolus*. Image: © Stuart Skeates.

Useful by-products of decomposer activity

When fungi grow in wood they often cause it to become stained, as a result of their coloured cell walls, coloured spores, changes associated with decomposition of wood, or chemicals produced during growth. While timber is usually devalued by such staining, because it is undesirable for manufacture of furniture or paper, colour changes by some fungi are greatly sought after. The beef steak fungus (*Fistulina hepatica*; page 37), a species that is fairly specific to oak and sweet chestnut, causes what is known as 'brown oak' heartwood which considerably increases the value of the timber for veneer production. This fungus decomposes wood very slowly so, unless it has been active for a very long time, the wood is still durable when felled. When the wood is dried the fungus ceases to grow. Green elf-cup (*Chlorociboria aeruginascens*) stains wood a blue-green colour, which was formerly much prized for use in the manufacture of Tunbridge ware – decorative wooden items inlaid with thin strips of coloured veneer forming patterns and pictures.

Decorative items are also made from what is known as spalted wood. Again, this effect is produced by fungi, this time as a result of one of three very interesting aspects of their biology. Firstly, a few fungi, including honey fungi (*Armillaria* species) and candle-snuff fungus (*Xylaria hypoxylon*), surround their decay columns in wood by what are called pseudosclerotial plates (PSPs). These appear as dark-coloured lines when wood is cut, but in three dimensions they form plates containing

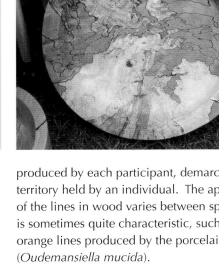

Far left and left:
Clearly visible dark lines
in spalted wood caused
by the formation of
pseudosclerotial plates.
Plates are commonly
formed at the junction
between the territories
held by two different
individual fungal mycelia
and are actually 'plates'
in three dimensions.
Images:
Box: © RBGE/Lynsey Wilson,
plate: © Lynne Boddy.

hyphae with many branches and large amounts of the pigment melanin, and often inflated or distorted into a variety of shapes. These PSPs insulate the decay column from the surrounding environment. In the case of honey fungus they prevent water loss causing the wood to become soggy (not surprisingly this wood is not used in woodwork!), while in candle-snuff fungus they keep the wood within dry. When wood is cut they are revealed as thin dark lines. The other two ways in which lines are produced are when (1) different individuals of the same species meet, and (2) mycelia of different species meet. These are called interaction zone lines and sometimes appear as double lines, one

produced by each participant, demarcating the territory held by an individual. The appearance of the lines in wood varies between species and is sometimes quite characteristic, such as the orange lines produced by the porcelain fungus (*Oudemansiella mucida*).

When mycelia of different species of basidiomycetes and of xylariaceous ascomycetes meet in soil, wood or other plant litter they interact aggressively (producing volatile and diffusible chemicals harmful to opponents) or parasitise opponents' hyphae, or both. The battle results either in deadlock, where neither species gains headway, or in replacement, where one species gradually takes over the

Left: Some fungi, such as
Armillaria gallica, create
patterns in partially decayed
wood. The patterns
arise from plates laid down
around columns of decay.
Wood with these patterns
is sometimes known as
spalted wood. When dried
the decay will cease so
manufactured items will last.
Image: © Peter Clarke.

Right: When the mycelia of two fungi meet in soil or dead organic matter they interact aggressively. Here mycelium of the stinkhorn fungus (*Phallus impudicus*) is overgrowing that of the sulphur tuft (*Hypholoma fasciculare*). Image: © Alaa Alawi.

Below: Fruit bodies of *Heterobasidion annosum*, the cause of Fomes rot that is a particular nuisance in conifer plantations, killing the trees and causing a white pocket rot of the heartwood. This fungus can be controlled by spraying conifer stumps with suspensions of spores of *Phlebiopsis gigantea* whose mycelium interacts aggressively with that of *H. annosum*. Image: © Alan R. Outen.

territory of the other; in some cases there is mutual replacement where fungus A starts to win the territory of fungus B in one region, but fungus B wins territory from fungus A in another region. Thus, fungal communities in organic matter change with time.

Some of these interactions can be used to man's advantage. Perhaps the best example is that of *Phlebiopsis gigantea*, which interferes with hyphae of *Heterobasidion annosum*. *Heterobasidion annosum* is a considerable nuisance in conifer plantations where it causes Fomes rot that kills and decays the roots and lower parts of trunks. When trees are felled, spores of *H. annosum* land on the cut surface of the stump and colonise the wood. It spreads to adjacent trees or seedlings through root contact, and can infect large areas. It can be controlled, however, by spraying suspensions of spores of *P. gigantea* onto the recently cut stump. These germinate and develop into mycelia which kill newly arriving *H. annosum* and any *H. annosum* that may already be present in the stump.

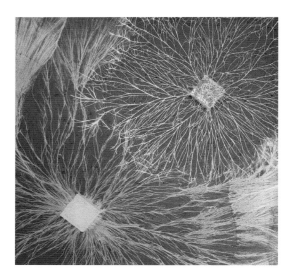

Climate change effects

Analysis of records of when fruit bodies have appeared since the mid 1950s shows clearly that fungal activity is changing in the UK and northern Europe. In the UK, decomposer fungi on average now start to fruit more than 40 days earlier in the autumn and continue fruiting much later (nearly 40 days) in the year.

Above: As a result of our changing climate some fungi now fruit in spring as well as in the autumn. This is because their mycelium is now active for much longer in the year. The wood-rotting sulphur tuft (*Hypholoma fasciculare*) is seen here fruiting outside the traditional autumn season. Image: © Martyn Ainsworth.

This does not mean, however, that every species now fruits earlier and carries on much later, rather that some species start to fruit later but continue for longer, whereas others fruit before these. Moreover, some fungi, such as the sulphur tuft (*Hypholoma fasciculare*), now fruit in the spring as well as in the autumn. These changes reflect climate change. They are not just of interest to those who go foraging for fungi (see Chapter 9) but indicate that fungi are active for longer. This means they are decomposing organic matter more rapidly. If this increase in decay simply matches the increase in plant matter produced as a result of climate change then there is no problem; but if decay is more rapid then there could be a problem as carbon from the store in the soil will be released to the atmosphere as carbon dioxide and will exacerbate the greenhouse effect. Crucially, this planet relies on decomposer fungi to decompose the most complex compounds found in plant materials. In so doing, fungi release nutrients for continued plant growth, and indirectly sustain animals too.

Coniochaeta polymegasperma species profile

Fungi that specialise in growing on dung are termed 'coprophiles'. To many people dung might seem an unpromising or unappealing place to look for fungi. Dung generally, however, has often yielded species previously unknown to science.

In the 1960s I first recorded an ascomycete that appeared to be *Coniochaeta hansenii* from the droppings of mountain hares (*Lepus timidus*) from various places in Scotland. As the spores were, however, significantly larger than normal my curiosity was pricked. In 1996 Roy Watling gave me some mountain hare pellets he had collected in Orkney on the island of Hoy, and the same larger spored *Coniochaeta* was present. With the passage of time, more specimens led me to the conclusion that this fungus was actually a new and quite distinct species.

The early 1990s coincided with my retirement, and I was able to devote more time to the study of coprophiles. The 'new species' turned out to be relatively frequent and widespread in Scotland, occurring on 21 of 45 samples from mountain hare. It was not recorded from other animal dung and initially, at least, was not recorded outside Scotland. Subsequently, I have recorded it from the Peak District of England and the Faroe Islands, but not from Ireland, Sweden or Finland. A reasonably thorough search of the dung of various other rabbit and hare species (northern and southern hemisphere, Europe and North America) has failed to turn up the new species. It is intriguing to wonder why *Coniochaeta polymegasperma*, as the species is now known, is apparently so restricted to mountain hares and, therefore, restricted in its own distribution.

Mike Richardson

Mountain hare. Image: © Lorne Gill/SNH.

Main photos, clockwise from this page:

Armillaria gallica;
Laccaria laccata;
porcelain fungus
(*Oudemansiella mucida*).
Images: © Peter Clarke.

Facing page inset photos:

Piloderma species
mycorrhizal mycelium (left).
Image: © Andy Taylor.

Rhytisma acerinum (right).
Image: © Chris Jeffree.

Chapter 3

Plant Pests and Perfect Partners

Lynne Boddy, Paul Dyer &
Stephan Helfer

Plant Pests and Perfect Partners

Lynne Boddy, Paul Dyer & Stephan Helfer

Right: About 90% of plants depend on fungi forming mutually beneficial associations with their roots – termed a mycorrhiza – to provide them with water and nutrients. Orchids, such as this heath spotted-orchid (*Dactylorhiza maculata*), have extremely small seeds containing only a small supply of food. Consequently, they must form this association as soon as their seeds germinate otherwise they will not survive. Image: © Peter Clarke.

Since plants first colonised land 450 or so million years ago they have been intimately associated with fungi. While the natural tendency is for people first to think of fungal foes in their gardens and fungal diseases killing food crops, the beneficial interactions between plants and fungi far outweigh the losses caused by the relatively few species that feed as parasites on plants. In fact, it is highly likely that it was fungi associated with primitive plants, before the evolution of roots, which allowed plants to move from water (where life began) onto land. There is absolutely no doubt that the ecosystems of planet Earth would not function without fungi. Nowadays, about 90% of plants depend on fungi forming mutually beneficial associations with their roots (termed a mycorrhiza – literally fungus root) for provision of water and vital nutrients. There is evidence of this type of association since plants first appeared in the fossil record.

Some fungi also form mutually beneficial associations with certain species of algae and cyanobacteria (what used to be called 'blue-green' algae) to form completely new organisms –

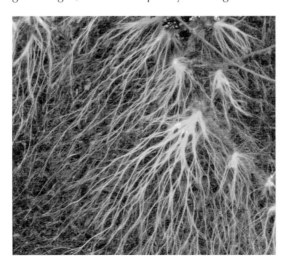

Right: Close-up view of the mutualistic mycorrhizal association between fungi and plant roots. The yellow threads are the mycelium of a *Piloderma* species. Image: © Andy Taylor.

lichens. Lichens are found worldwide and are the dominant form of vegetation in some ecosystems. In lichens the algae are found enclosed by fungal hyphae. A contrasting situation also occurs in which fungi are completely enclosed within plant tissues. This includes not only fungal pathogens of plants but also endophytes (literally meaning 'within plants') that are harmless or even beneficial to plants.

This chapter takes a 'whistle-stop tour' of interactions between plants and fungi, ranging from the diseases that cause damage and destruction and which have dramatically shaped our landscape and history in the past, through to a variety of beneficial relationships seen in mycorrhiza, lichens and endophytes. Finally, we consider how global change is affecting these relationships, and what we might do to maintain an appropriate balance between fungi and plants in the natural world.

Plant pests
The extent of the problem

Imagine if you could not have any more coffee or that your country's staple crop failed repeatedly and there was no substitute. You may think that these are questions that only arise in developing countries, but they are problems experienced in the British Isles relatively recently. Both a rust fungus and a fungus-like downy mildew have caused massive crop failure with profound consequences.

Around 1869 the coffee rust fungus (*Hemileia vastatrix*) invaded the coffee plantations of Ceylon (now Sri Lanka), making the cultivation of coffee for the British market unprofitable. The alternative crop grown ever since, of course, has been tea, hence the British liking for it.

In the early 1840s Ireland and other countries were highly dependent on potatoes for feeding large numbers of their populations. Because potatoes are propagated by tubers much of the crop was genetically uniform (just a few clones in cultivation). When a new pathogen arrived in 1844 – the fungus-like late blight (*Phytophthora infestans*) – it had no difficulty colonising a large proportion of the potato crop, and weather conditions resulted in it causing unprecedented loss. Similar conditions occurred in the following years. As a consequence one million people in Ireland died of starvation and a further million emigrated, mostly to North America.

It is estimated that worldwide, the cost of average annual crop losses is about US$550 billion. Of these, 60% are due to weeds, animal pests, droughts, frosts and other

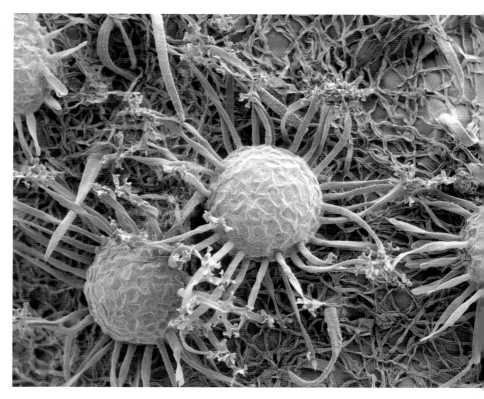

Above: Microscopic fruit bodies of the powdery mildew *Erysiphe azaleae*, a parasite of rhododendrons, as seen with a scanning electron microscope. Image: © RBGE/Stephan Helfer.

environmental factors; the remaining 40% are due to plant diseases, of which around two-thirds are caused by fungi (e.g. the cause of coffee rust) and fungus-like organisms such as late blight, and the rest by other microorganisms. Without crop protection the potential loss would be more than double.

Which fungi cause disease?

All members of the fungus-like organisms called downy mildews (*Oomycota*; see Chapter 1) are parasites. They are responsible for damage to food and forest plants as well as ornamental shrubs and trees. One genus, *Phytophthora*, has around 200 known species (with an estimated 200 more as yet undescribed) and is particularly widespread, having a huge range of host plants. *Phytophthora infestans* has already been mentioned, and *P. ramorum* is currently causing sudden oak death. Many other members of the *Oomycota* cause serious plant disease, ranging from damping off of seedlings to root rots.

Within the sac fungi (*Ascomycota*; see Chapter 1) there is an entirely parasitic section – the powdery mildews. Many gardeners will recognise mildew on roses and rhododendrons, peas and peaches. Their main economic

Left: Late blight of potatoes became such a catastrophic problem largely because the potatoes that were being cultivated in the early 1840s were genetically uniform. This photo shows the wide genetic diversity that is vital for breeding improved crop varieties. Image: Emanuele Biggi/OSF/ www.photolibrary.com

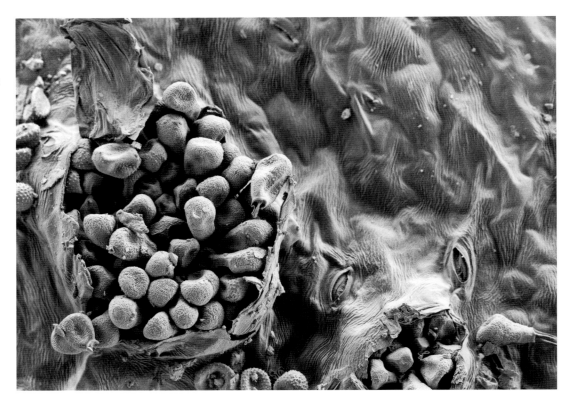

damage is on cereal crops, especially wheat and barley, where *Blumeria graminis* inflicts heavy losses every year. Dutch elm disease (*Ophiostoma novo-ulmi*), ergot (*Claviceps*), scab (*Venturia*) and leaf roll (*Taphrina*) are other examples of parasitic fungi.

Alongside these, and equal if not exceeding in impact, are the parasitic species belonging to the club fungi (*Basidiomycota*; see Chapter 1). Being very specific in their host range, these parasites live inside their hosts and cause damage in a wide variety of ways. The rusts are serious crop and wild plant pathogens. With in excess of 5,000 species, they attack ferns, conifers and flowering plants, and are responsible for a 10–15% loss of cereal production on average. In epidemic years this can amount to 50% in crops without protection. Smut fungi similarly affect many crop plants, causing the whole of the seed to be changed into fungal spore mass (e.g. *Ustilago maydis*). Control is therefore possible using seed hygiene and treatment methods. Other parasitic members of the club fungi are the leaf gall (*Exobasidium*), root and butt rot-causing fungi such as *Heterobasidion annosum* (causes Fomes rot; see Chapter 2) and honey fungi (*Armillaria* species). *Heterobasidion annosum* can cause huge losses in conifer plantations by spreading from tree to tree via root grafts. Some *Armillaria* species, on the other hand, spread from tree to tree via root-like rhizomorphs, killing trees over many hectares.

There are many *Armillaria* species worldwide and probably seven in Europe. The most common in Britain are *A. mellea*, *A. ostoyae* and *A. gallica*. The first two are

splitting of bark. Cankers are sometimes formed with exudates under loose bark, and plants die. Of course, other diseases can produce similar symptoms to these. Confirmation that the culprit is an *Armillaria* species can be obtained by removing a small area of bark that might reveal the white mushroom-smelling sheet of mycelium and flattened brown rhizomorphs that these species produce. Rhizomorphs may be seen in the soil and in autumn large clumps of honey-coloured fruit bodies often appear close to the trunk or growing along the ground following a root.

Control is difficult; infected wood and roots should be removed, and plants should be well tended to try to maintain them in a healthy and vigorous state. In badly infected gardens the only recourse may be to plant less susceptible varieties or move house!

This may paint a rather worrying picture, but most plants are immune to most plant pathogens. Many of our problems are caused by planting clones (selected for high yield) that allows pathogens to rampage through the whole crop, once they become established.

Left: There are several species of honey fungus (*Armillaria*), some of which are devastating tree pathogens while others are saprotrophs feeding off dead wood or weakened trees. When colonising trees, *Armillaria* is often seen growing beneath the bark as rhizomorphs (root-like organs). Image: © Patrick Hickey.

Below: The fruit bodies of different honey fungi (*Armillaria*) look very similar. Distinguishing between species is very important from a gardener's or forester's point of view, as not all are pathogens. Image: © Patrick Hickey.

plant pathogens and obtain their food by killing plants via their roots, while the latter feeds on dead plant tissues, but may attack suppressed and weakened plants. *Armillaria mellea* is a common and often devastating pathogen of a large range of forest trees, garden trees and ornamental and herbaceous plants, whereas *A. ostoyae* is most common on conifers. *Armillaria gallica* is most common on broad leaved trees.

The fruit bodies of different honey fungi look very similar. Consequently, it was not realised that there were several species having different hosts and ecology until the early 1980s. From a gardener's or forester's viewpoint, distinguishing between species is important but not straightforward. They all have a ring present on the stem. With *A. gallica* the stem is often swollen and yellow. Fear often strikes the heart of gardeners if they spot the root-like 'bootlaces' typical of many species, but this may be an overreaction as *A. mellea* is not a prolific producer of rhizomorphs though *A. ostoyae* and *A. gallica* are.

Symptoms of infection include colouring of leaves, which then fall prematurely, covering of resin and mycelium on conifer roots, and the

Right: Cluster cup pustules of the rust fungus *Puccinia urticata* growing on the leaf stalk of nettle.
Image: © Patrick Hickey.

How do parasitic fungi spread?

Spread via rhizomorphs of honey fungi and tree root contact by *Heterobasidion* is actually unusual. Most parasites are spread as spores, that can be carried in water, in the wind or by animals. For example, mildew spores may get onto passing animals or coats of humans. By far the greatest agent in the long-distance spread of new diseases in recent times has been the shipment of diseased plants or plant products and international travel: Dutch elm disease arrived in Europe in the 1910s on infected timber from the Himalayas, was shipped to North America in the 1920s on wood for veneer, and returned to Europe, transformed and more aggressive, in infected logs in the 1960s. Sudden oak death (*Phytophthora ramorum*) arrived at the Royal Botanic Garden Edinburgh's Regional Garden at Benmore on the shoes of visitors, who had previously visited infected sites elsewhere (the exact source is not yet known).

Once a spore has arrived on a host plant, the parasitic fungus must get through the protective outer layers into the tissues within the plant, to extract the nutrients it requires for living. There are various routes of entry. Rusts, downy mildews and some other parasites enter through the natural openings in the leaves, the stomatal pores. Other parasites, such as powdery mildew fungi, penetrate directly into the outer plant cells, using specific enzymes

Below: Plant pathogenic fungi enter plants in different ways. Some enter through wounds, others force their way directly through the cell walls, and others enter via natural openings. Hyphae of *Puccinia hordei*, the cause of barley brown rust, grow across the surface of a barley leaf searching for stomata (gas exchange vents in the leaf surface) through which they can gain entry to the inside of the leaf.
Image: © Nick Read.

to breach the host plant cell wall. Other fungi produce toxins which disrupt the host plant's metabolism, whilst yet others rely on wounds to invade the plant.

After successful penetration has occurred, fungi may grow between the host plant's cells, sending feeder hyphae into the host plant's tissue in order to assimilate their nutrients. In the process an intimate relationship arises between host and parasite, somewhat similar to that in mycorrhizal symbiosis (described below), except that the plant does not receive any benefit from the parasite! As with mycorrhizas, the fungus is dependent on the living host, and killing it would spell its own demise. This non-lethal feeding is called biotrophy – eating the living. In other, less specialised relationships, the fungi may produce harmful substances, killing the host cells and living on the resulting dead tissue. This is called necrotrophy – eating the dead.

Plants do defend themselves, and responses range from abandoning whole organs (e.g. leaf drop), to actively sacrificing cells surrounding an infection site ('hypersensitive' response) to inhibiting the parasite's ability to produce spores. Dark phenolic compounds can often be seen in woody hosts defending themselves against fungal invasion. The attack and defence responses have evolved over thousands of years, and in some cases parasites are extremely specific for the host that they can colonise. For example, specific strains of *Puccinia graminis* can only parasitise certain varieties of certain species of cereals.

Control of disease and weeds

From the beginning of agriculture, farmers saw what would harm or benefit their crops, and so it is not surprising that disease control was discovered (and subsequently forgotten) by the ancient farming communities in the Fertile Crescent, where sulphur was used to treat affected cereal crops against disease. This chemical is still in use as a fungicide. Other traditional methods of control, though not often used consciously, include selection of successful varieties, crop rotation, removal of carrier hosts, crop hygiene and physical (heat) treatment. Modern control methods rely on the selective discouragement of the pathogen without harm to the host and can involve the breeding of disease-resistant varieties, chemical control, seed treatment, physical barriers and quarantine, the use of biological antagonists against the parasite and genetic modification (GM).

On the flipside, plant diseases can be used in the fight against unwanted plants such as invasive weeds. An example is the successful treatment of skeleton weed in the USA using the rust fungus *Puccinia chondrillina* as a bioherbicide. Rust fungi are particularly suited because they are very host specific, and non-target plants are not at risk.

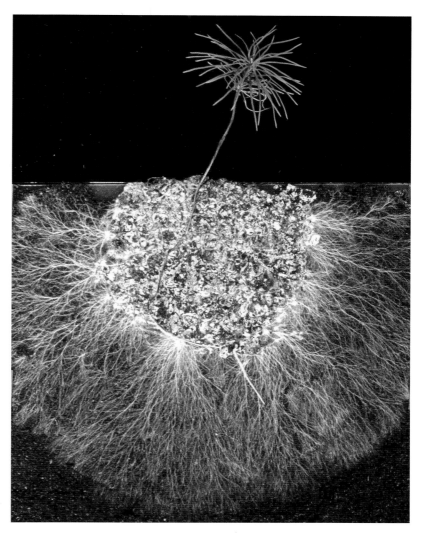

Driving life on planet Earth – mycorrhizas

As already mentioned, the vast majority of plants in nature have intimate associations between their roots and fungi. There are an estimated 400,000 species of plants on the planet and probably at least as many species of fungi intimately associated with their roots, so it comes as no surprise that there is a wide range of different types. Only the most common and most intriguing are described here. In all types fungal hyphae are found within young, active roots growing between or within plant cells. In some types hyphae completely sheath the roots. In all types hyphae spread away from roots into soil, and sometimes they link up with mycorrhizal mycelium from other plants, forming a bridge between different plant species. This network is sometimes called

Above: Forest trees all depend on fungi forming mutually beneficial associations with their roots – mycorrhizas. There are different types of mycorrhizas, the most common on trees being called ectomycorrhizas. On this pine seedling most of what is visible are the hyphal filaments of the brown roll rim (*Suillus bovinus*) extending from those roots into soil in search of water and nutrients. Image: © Jonathan Leake.

Left: Many countries take insufficient precautions to prevent the spread of plant pathogens. Pathogens can be imported when plants enter a country. Some pathogens can be spread on humans. To try to prevent the spread of the fungus-like pathogen *Phytophthora ramorum*, visitors to RBGE walk on mats of disinfectant to kill the spores. Image: © RBGE/Vlasta Jamnický.

Clearly, these fine hyphae form a huge surface area that can absorb water and mineral nutrients from soil, and are capable of exploring for nutrients much more extensively than roots. Thus, although we are taught from an early age that it is roots that obtain water and nutrients for plants to grow, in the vast majority of cases in nature it is fungi that primarily fulfil this role. The mycelium in soil and the sheath around the root, in those types where this is produced, can also protect roots from soil-borne plant pathogens. In addition, mycorrhizal fungi can protect plants from toxic chemicals in the soil, and plants with appropriate mycorrhizal partners can be used to reclaim polluted land (e.g. *Pisolithus tinctorius* tolerant of acidity in coal spoils in the USA). Fungi do not provide a 'free service' and plants pass about 18–20% of the sugars made during photosynthesis to the fungus. Exchange of water, nutrients and sugars occurs between hyphae and plant cells within the roots.

It is impossible to overstate the importance of mycorrhizal fungi since without them most plants would not grow successfully, and without plants there would be no food for animal life, including humans!

Arbuscular mycorrhizas (AM) are the most common and widespread type, forming with herbaceous plants and some trees, such as maples, ferns and occasionally mosses and liverworts, and only a few plant families lack them, such as the cabbage family (Brassicaceae) and the goosefoot family (Chenopodiaceae). They are an ancient association and, as already mentioned, have been found fossilised. Within plant roots they grow between the cortex cells and also grow into cells, where they branch profusely, forming arbuscles – literally dwarf tree. These arbuscles form a large surface area and are the sites of nutrient transfer to the plant. Hyphae extend into soil from root epidermis cells and from root hairs. The fungi involved belong to the *Glomales*. They do not produce macroscopic fruit bodies, but spread by producing spores from the hyphae in soil.

Sheathing mycorrhizas (also called ectomycorrhizas) are the second most common type overall, and the most common type on trees. Essentially they are made up of a sheath of fungal material surrounding the fine roots,

Above: The most common mutualistic relationships between plant roots and fungi are called arbuscular mycorrhizas. The bluebell is one plant which forms this type of association. Image: © RBGE/Max Coleman.

'the wood-wide web'. The amount of fungal mycelium in soil is enormous and difficult to estimate, but it has been claimed that there are about 2.2 metric tons of fungi in soil for every person on Earth. The surface area of soil fungi has been estimated as 2–9 times that of the Earth, and the soil beneath a single square metre has been estimated to contain 16,000 km of mycelium.

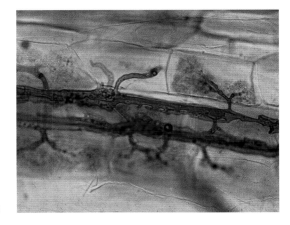

Right: Species of *Glomus* commonly act as arbuscular mycorrhizal partners. This section through a plant root shows *Glomus* species hyphae and arbuscles (literally meaning dwarf tree) in plant cells. Image: © James Merryweather.

from which some hyphae extend into soil and others into the roots, between but not penetrating the cortex cells to form a network termed the Hartig net, after the German biologist Robert Hartig. Exchange of water, mineral nutrients and sugars occurs in the Hartig net. Trees form mycorrhizas with more than 5,000 different species of fungi, mostly basidiomycetes, for example species of *Russula*, *Boletus* and *Amanita*. Some fungi form mycorrhizas with several tree species, whereas others are more specific, for example *Suillus grevillei* with larch, and *Leccinum pseudoscabrum* with hornbeam. Although some tree species associate with only a few fungal species, such as red alder only with *Alpova diplophloeus* and *Lactarius obscuratus* in the USA, most tree species host a range of fungal species. Indeed, a single tree can form mycorrhizas with many species at the same time, with different fungi associated with different roots, and with different species during a tree's long life. New roots are typically colonised by hyphae in soil that are already attached to a host tree, linking different roots on a host tree and different trees. The sheathing mycorrhizal fungi produce visible fruit bodies, and the group includes many familiar larger fungi. Some, like Satan's bolete (*Boletus satanas*), royal bolete (*Boletus regius*) and the stipitate

hydnoids (stalked tooth fungi) *Hydnellum*, *Phellodon*, *Sarcodon* and *Bankera*, are rare and endangered.

The basis of the mycorrhizal mutualism is that plants provide sugars and fungi provide water and nutrients (and protection from some pathogens). However, some fungi have managed to alter the balance in their favour and have become plant cheaters, whilst in other cases the plants have the upper hand. Orchids have very small seeds and upon germination in the natural environment they must quickly form a mycorrhizal association otherwise they will not survive. The fungal partners not only supply water and minerals but also sugars until the orchids are able to photosynthesise. The fungi have to obtain these sugars from other sources. Some are wood decayers (e.g. species in the genera *Trametes* and *Marasmius*) whilst others are plant pathogens (e.g. species in *Armillaria* and *Rhizoctonia*). The latter are 'dangerous allies' as they will actually kill the orchids if they do not keep their part of the bargain by supplying sugars. Some orchids lack chlorophyll for their whole life, so are unable to photosynthesise, and hence are entirely dependent on fungi. Effectively these orchids parasitise the fungi for water and nutrients, and other plants for sugars using the fungus like an umbilical cord.

Left: The non-photosynthetic bird's-nest orchid (*Neottia nidus-avis*) is unable to produce its own sugars. The plant is entirely dependent on mycorrhizal fungi not only for water and mineral nutrients but also for energy. Effectively it is parasitic on the fungus. Image: © Stuart Skeates.

Below: Fruit bodies of the fly agaric, *Amanita muscaria*, a species that forms ectomycorrhizas with a number of tree species, including birch. Image: © Patrick Hickey.

Right: Indian pipes (*Monotropa uniflora*) has lost its ability to make chlorophyll, so cannot photosynthesise. It obtains its water and nutrients from a fungus partner, and its sugars from the surrounding trees via the fungus link. Image: © Andy Taylor.

Other plants are similar 'cheaters' in that they also form mycorrhizas but do not photosynthesise and hence can feed neither themselves nor their fungal partner. An example of this is the yellow bird's-nest (*Monotropa hypopitys*), that grows in conifer forests. The plants have fleshy root balls, and the fungi within form a Hartig net and mantle, like the ectomycorrhizas. Indeed, the fungi involved, including species of *Russula*, *Rhizopogon* and *Suillus*, all form ectomycorrhizas with forest trees, and sugars flow from the host tree to the parasitic plant.

Surviving extreme environments

In the same way that certain fungal species form mycorrhizas to the mutual benefit of the fungus and plant roots, other specialised fungi are able to form intimate relationships with green algae and/or cyanobacteria to form whole new symbiotic organisms – the 'lichens'. This is an extremely important way of life for these fungi, with around 15–20% of all known fungal species forming lichen associations. These fungi are almost entirely ascomycetes (some 40% of all known ascomycete species), though about 2% of lichens are formed by basidiomycete fungi. In contrast, only a limited number of algal and cyanobacterial species form lichens, with the most common genera being the green algae *Trebouxia* and *Trentepohlia*, and the cyanobacterium *Nostoc*. In lichens the mutually beneficial relationship involves the fungal 'mycobiont' partner capturing water and mineral nutrients from the surrounding environment, which it then provides to the algal or cyanobacterial 'photobiont' partner in exchange for sugars produced during photosynthesis. This sugar provides the fungus with the energy it needs to grow. Some cyanobacteria also fix atmospheric nitrogen, which is passed on to

Right: Lichens are composed of a fungus plus algal or cyanobacterial cells. The fungus captures water and nutrients and in return the photosynthesising partner provides sugars. Different lichen species have very different growth forms. The lichen *Ramalina menziesii*, draped across tree branches in the Pinnacles National Park, California, USA, has a fruticose form. Image: © Paul Dyer.

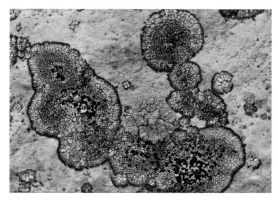

Far left: *Buellia frigida* (black) and *Caloplaca saxicola* (orange) are crust-forming (crustose) lichens. They are seen here growing on rock in the Rothera region in Antarctica. Image: © Peter Crittenden.

Left: *Usnea subfloridana* is a bushy (fruticose) lichen, seen here on birch in Raasay, Scotland. Image: © Mike Sutcliffe.

the mycobiont to assist growth. The fungus also forms a hyphal matrix, which may be highly pigmented, to protect the photobiont from adverse environmental conditions and excessive solar radiation. However, some scientists have argued that the lichen symbiosis can be viewed as a type of controlled parasitism by the fungi.

A key feature of the lichen symbiosis is that the pattern of growth of the fungus, alga or cyanobacterium alone changes markedly when the two partners come together. A lichen's form is unique to the particular symbiotic partners involved. Lichen-forming fungi have only very limited growth ability without their photosynthetic partner, an indication of their highly specialised nature. A variety of different lichen growth forms can be recognised, such as 'crustose' lichens that grow as an adhering thin crust on surfaces, 'foliose' lichens that appear leaf-like or with lobes, and 'fruticose' lichens that have a bushy or shrubby appearance. These lichens can exhibit a fascinating diversity of colours, ranging from dark black through to bright orange, red, yellow and white. A cross section through the lichen reveals the photobiont partner intermingled between the fungal hyphae, with the mycobiont being the dominant partner, forming 80% or more of the biomass. Lichens range in size from typically less than 10 cm in diameter to some fruticose forms that can be over a metre in length. It has been suggested that the manner of growth of some fruticose *Usnea* species draped over tree branches led to the custom of putting tinsel on Christmas trees! However, lichens in general have very slow growth rates, rarely exceeding 10 mm per year.

Lichens are able to reproduce and disperse by various methods. The majority of lichens produce sexual ascospores from the fungal partner that are wind or animal dispersed and must meet a photobiont partner on a suitable substrate to re-establish the lichen symbiosis – a rather chancy affair. Some lichens also release small fragments called 'soredia' and 'isidia'. These contain cells of both symbiotic partners parcelled up together so that after dispersal the fragments can begin growth immediately.

Lichens are arguably one of the most successful types of organism in the world. They are found in terrestrial environments globally, ranging from the polar regions of the Antarctic and Arctic, through to temperate

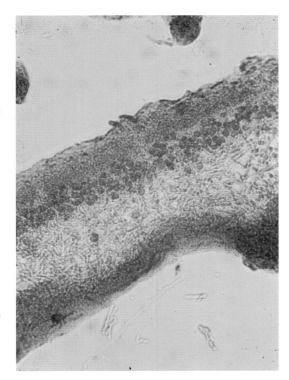

Left: Cross section through a lichen showing cells of the algal photobiont partner (in green) present within the interior of the lichen sandwiched between the upper and lower fungal (mycobiont) cortex and medulla regions. Image: © Peter Crittenden.

Right: Rock, stone monuments and gravestones are undisturbed environments that are stressful for most plants because they are low in nutrients, often very dry and exposed. They are usually colonised by certain species of lichen, which are adapted to these conditions, as seen here at Arisaig on the west coast of Scotland. Image: © RBGE/Max Coleman.

Below: Lichens are important components of vegetation at high latitudes in northern boreal forests and Arctic tundra as illustrated here in spruce lichen woodland close to the forest treeline in Northern Quebec, Canada, where the ground vegetation is dominated by *Cladonia* reindeer lichens. Image: © Peter Crittenden.

and tropical regions and even deserts. A key reason for their success is that lichens are able to survive and grow in environments subject to poor nutrient and water availability and extremes of temperature, where few other organisms can survive. This includes growth on rock surfaces, nutrient-poor soils, tree bark and twigs, and even within rocks in the Antarctic. Lichens are also a familiar feature of urban areas, seen growing on gravestones, roof tiles, paving slabs and trees in temperate countries. In some terrestrial ecosystems, such as the Arctic tundra and boreal forests of Russia and Canada, lichens dominate. Here, 'reindeer' lichens can form extensive mats covering the soil surface. It is thought that the most dominant fungal species worldwide in terms of total biomass is one of these *Cladonia* reindeer lichens. The sheer quantity of these lichens is sufficient to support large herds of reindeer and the human communities that rely on them.

The success of lichens stems from the fact that both symbiotic partners can work together to survive under conditions which neither partner alone could tolerate. Lichen-forming fungi are especially good at sequestering mineral nutrients from atmospheric depositions such as rain, snow melt or animal urine or excreta. Growth normally occurs during brief

windows of opportunity when conditions are favourable, such as following rainfall, in the brief polar summers or during early morning dewfall or fog in deserts. At other times the lichen can remain dormant in a desiccated state, in which they are remarkably resistant to adverse conditions. For example, desiccated lichens have been exposed to the vacuum and solar radiation of outer space on orbiting space satellites; amazingly, when they were recovered after landing on Earth, they were able to resume growth following rehydration. Some even suggest that such life forms might be transmitted from planet to planet!

As well as being a key component of global biodiversity, lichens also have important ecological roles. Lichens are key pioneer colonisers of bare rocks and infertile soils and begin the process of soil formation and vegetation succession. Carbon fixation by lichens in polar regions may represent a significant energy input in the absence of other dominant vegetation types. Due to their ability to capture low levels of nutrients lichens can be used as biomonitors of pollution levels, because many lichens are inherently sensitive to atmospheric pollutants. Other lichens, however, are more pollution tolerant and levels of atmospheric pollution can be predicted from the observed lichen flora – a nice school exercise (see OPAL link in bibliography).

Lichens also have some economic value. They have been used in the production of perfumes, cosmetics and dyes, given insights

into mineral prospecting, and are used in wreaths and as model trees in miniature landscapes. Lichens also produce intriguing metabolites called 'lichen substances', which might have novel medical uses.

The hidden world of endophytes

Some fungi – termed endophytes – spend all or parts of their lives within plant tissues, but do not cause any symptoms of disease. They are found in most living plant tissues, including seeds, roots, shoots and leaves, from tropical mangroves to roots of polar and alpine plants, and from grasses to trees. Some can be thought of as 'accidental tourists' while others are 'influential passengers'. Some probably confer resistance against pathogens and insect pests, and deter large grazing animals, as a result of chemicals they produce. Some of these chemicals, such as the anti-cancer agent taxol, first isolated from yew, have medicinal uses for humans, and no doubt many novel chemicals produced by endophytes await discovery (see Chapter 5). The relationships between plants and their fungal endophytes are still not clearly understood, but it is known that the local environment can change the balance of the relationship, so that a benign endophyte, which had been residing in host tissues without symptoms in the host, can sometimes turn into a pathogen. Climate change is likely to affect the

Above: Some fungi (endophytes) that live within plant tissues without causing disease symptoms produce chemicals that benefit the plant and man. Some confer protection against plant pathogens and insect pests. Some from yew, and elsewhere, produce the anti-cancer agent taxol. Image: © RBGE/Lynsey Wilson.

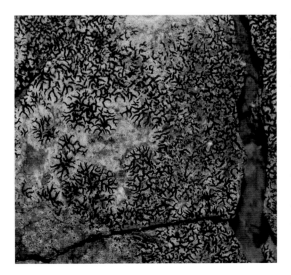

Left: Some lichens, such as the script lichen *Opegrapha vulgata* are sensitive to air pollution and can be used as biomonitors. It is even possible to reconstruct past environmental conditions from lichens preserved on the internal timbers of old buildings. Image: © Mike Sutcliffe.

Above: Lichens, such as this *Ochrolechia* species growing on rock at Mount Tumbledown, the Falkland Islands, are likely to be particularly affected by climate change since their environments are already extreme. Image: © Peter Crittenden.

Impressive manifestations of these hidden fungi are sometimes seen on tree trunks following drought. After the 1976, 1977 and 1990 droughts in Britain, long thin (5–10 cm) dark strips were seen spiralling around the trunks of beech trees. At a quick glance these looked rather like lightning strikes, but closer inspection revealed they were in fact strip cankers – dead bark and underlying tissues. The fungi that caused the strip cankers were present in the water-conducting sapwood as sparsely dispersed spores, yeasts or hyphae. During the drought some major roots were unable to obtain sufficient water and the columns of wood associated with these began to dry out. The latent fungi were then able to develop and formed the extensive strip cankers in a matter of weeks.

presence, distribution and activity of endophytes within plant tissues, and experiments growing plants in air with increased levels of carbon dioxide have revealed different distributions of endophytes. As with all fungi, endophytes must be able to spread from one plant to another. Some do this as spores spread by wind and/ or insects. Others spread from one plant generation to the next by fungal hyphae that have penetrated the host's seeds.

A changing world

Man's activities have had an increasing impact on the world's habitats and climate. These environmental changes impact on fungi and on their plant partners in many ways. For example, the time when mycorrhizal fungi fruit has changed since the late 1970s, probably in response to climate change (see Chapter 10). Climate change may also affect which fungi form mycorrhizas. Nitrogen pollution of soils

Right: Our changing world is affecting fungi. Some mycorrhizal fungi now fruit at different times because of changing climate and some have become quite rare because of nitrogen pollution. *Phellodon melaleucus,* here fruiting in mixed deciduous woodland, has been severely affected by nitrogen pollution. Image: © Martyn Ainsworth.

has certainly had major effects on mycorrhizal fungi; species in the genera *Hydnellum* and *Phellodon* are now very rare, though in the Netherlands they are increasing now that nitrogen pollution is declining.

The incidence and severity of plant diseases is likely to alter considerably; some diseases will become more widespread and more severe, while others may be more restricted or less damaging, and the timing of diseases may change. The geographical range of many plant pathogens is determined by whether or not they can survive the winter. For example, warmer winter temperatures increase the survival of the rust fungus *Puccinia graminis*, which leads to increased disease of grasses and cereals. The geographic range of the fungus-like *Phytophthora cinnamomi* in Europe is expanding in response to increased temperatures, allowing the species to overwinter further north and at higher altitudes. This pathogen causes root rot and dieback of forest trees, fruit trees and a wide range of ornamental plants. On the other hand, higher temperatures lead to better and longer seasons of plant growth. Perhaps even more worrying than climate change is the influence of global movement of plants and goods, and international travel, which introduces pathogens to new areas. To secure our food supplies and to minimise plant losses we need to take measures to try to prevent arrival of new pathogens, to adapt cultivation methods, and to understand the biology and ecology of plant pathogens.

Lichens are likely to be particularly affected by global change because they are already found in extreme environments, where small changes in conditions could prevent survival altogether. For example, increased radiation levels due to the ozone hole in polar regions might be detrimental for the growth and survival of lichens in these areas. Increasing urbanisation and deforestation is leading to habitat loss for endangered tree-dwelling lichens. Pollution is also a concern (see Chapter 10). Some lichen species cope well whereas others do not. What the knock-on effects of such changes will be is hard to predict.

Cladonia stellaris (reindeer lichen) species profile

Cladonia stellaris is one of several species of ground-dwelling lichen that form spectacular carpets in subarctic forests and tundra. These lichens are a major component of the winter diets of caribou in North America and domesticated reindeer in Eurasia, giving them their common name of 'reindeer lichens'. *Cladonia stellaris* forms carpets more than 15 cm deep across large areas. Bearing in mind that there are over four million square kilometres of lichen-dominated forests in Canada alone, *C. stellaris* is a likely candidate for one of the world's most abundant fungi. It is also food for thought that the symbiotic algae in the lichen are one of the principal photosynthetic organisms over these large areas, and as such have a significant role in carbon dioxide capture and oxygen production.

Lichens are renowned for their slow growth but reindeer lichens grow comparatively rapidly. Reindeer lichens decay at their bases and so are not firmly attached to the underlying soil. This is relatively unusual amongst lichens, the majority of which are firmly attached to surfaces such as rock, tree bark or soil. It is thought that the decay of the older parts might contribute to the rapid growth via internal recycling of nitrogen and phosphorus – important nutrients that are scarce in the habitats in which these lichens are so successful.

The often deep layer of dead lichen that accumulates beneath the *Cladonia* mats also appears to insulate the lichen from the chemical influence of the underlying soil; this makes reindeer lichens particularly good indicators of atmospheric chemistry.

The fine multi-branched structure of reindeer lichens makes them fragile and prone to damage from trampling. Walking over lichen-covered ground in summer when the lichen mats are dry will leave deep boot prints. At this time of year caribou and reindeer are further north grazing the tundra and so do not trample the lichen. Instead, the animals inhabit these lichen-rich habitats in winter when snow cover of one metre or more protects the lichen carpets from trampling. The caribou and reindeer can smell lichen through such depths of snow and dig craters in the snow to feed.

Peter Crittenden

Image: © Peter Crittenden.

Main photos, clockwise from this page:

Lycoperdon perlatum.
Image: © Peter Clarke.

Trametes gibbosa.
Image: © Chris Jeffree.

Termites cultivating a fungus comb of the fungus *Termitomyces reticulatus.*
Image: © Karen Machielsen.

Facing page inset photos:

Fruit body of a new *Ophiocordyceps* species from Ghana (left).
Image: © Harry Evans.

Toads killed by *Batrachochytrium dendrobatidis* (right).
Image: © Matthew Fisher.

Chapter 4

Animal Slayers, Saviours and Socialists

Harry C. Evans & Lynne Boddy

Animal Slayers, Saviours and Socialists

Harry C. Evans & Lynne Boddy

Fungi interact with and impact upon animals both directly and indirectly in many ways. We interpret these associations as being positive or negative in relation to humans but this is often too simplistic, as well as ambiguous. For instance, the white muscardine fungus (*Beauveria bassiana*) attacks commercial silkworm cultures, and is seen as a pest, but in nature this species may contribute to controlling silkworm populations and prevent mass defoliation of the insect's preferred food plant – mulberry trees. Taking an ecological view, rather than an economic one, the role of the silkworm pathogens may be considered beneficial to the functioning of the ecosystem. The delicate balance of nature is a theme we will return to later.

A small number of fungi can impact directly on vertebrates by colonising and growing on or within the skin or inside the body. Until recently, such infections were considered to be an inconvenience rather than a significant health problem for humans. Dandruff is caused by *Malassezia* yeast. Though troublesome, the ringworm or tinea group of diseases, which include athlete's foot, are not serious infections. Similarly, although fungal spores are ever present in the air we breathe they have not been viewed as a particularly serious threat to human health. However, some fungi are allergens, causing responses such as hay fever, and are sometimes more problematic. A few fungi can cause major diseases in humans (see below). In most places, including the UK, major illnesses caused by fungi are relatively rare, though fungi are often ultimately fatal to patients whose immune system is already severely compromised.

In sharp contrast, the interactions of fungi with invertebrates, though sometimes subtle, have far greater global significance. These interactions are the result of millions of years of coevolution and are finely tuned and highly specific. Some impact negatively on animal populations, while others form positive or mutually beneficial associations. Such associations are especially developed in social insects, particularly termites and ants, and have allowed them to achieve key roles in many ecosystems.

Below: White muscardine fungus (in this case, an unnamed species of *Beauveria*) on a moth chrysalis hanging in forest canopy. Another member of this genus, *B. bassiana*, attacks commercial silkworms. Image: © Harry Evans.

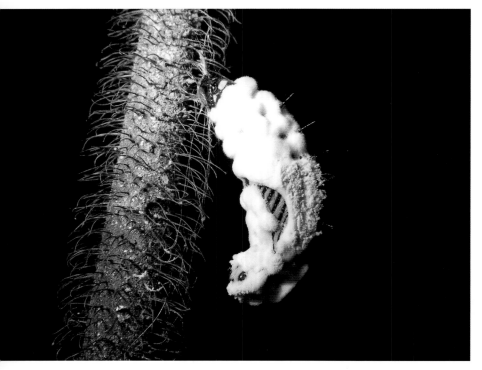

Invertebrate slayers and saviours

Fungi are major killers of invertebrates, and thus play an important part in determining their population dynamics. The idea of exploiting these pathogens to control invertebrate pests – from slayers to saviours – has met with fluctuating fortunes over the past century or so. Difficulties in manufacturing biopesticides and inconsistent field performance meant that they were unable to compete with chemical pesticides. However, improved technology, as well as increasing concerns over the misuse and long-term dangers of applying chemicals, could lead to these products becoming more mainstream in the battle against invertebrate pests.

Before exploring fungal control of invertebrate pests further let us first consider the forests of the humid tropics where these fungi evolved. These forests can help answer many questions.

What species diversity do we find? What are the logistics and mechanisms involved in the fungi reaching their host targets? What weapons do they deploy to slay their victims? Above all the study of natural habitats reveals their key but cryptic role in ecosystem functioning.

The tropical primary forest is a delicately balanced ecosystem in which all of its organisms are interdependent. For example, there is evidence that the high diversity of tree species is, in part at least, determined by host-specific fungal pathogens, and there is every reason to believe that invertebrate pathogens exert similar evolutionary pressures. However, since the majority of invertebrates are themselves microscopic, the true scale of the diversity and impact of fungal pathogens on animals in these, and indeed all, terrestrial ecosystems can only be imagined.

Above: View of a tropical forest – one of the most biologically diverse ecosystems on the planet and a treasure chest of entomopathogenic fungi. Image: © RBGE/Peter Wilkie.

Right: Some fungal pathogens cause mummification of their victim. Here the mycelium of *Beauveria bassiana* is just beginning to emerge from the spiracles of a mummified caterpillar. Image: © Harry Evans.

Far right: The same caterpillar four days later. The white surface comprises fungal filaments (hyphae) and spores. Image: © Harry Evans.

Below: Sphingid moth attached to tree bark bearing numerous white, cylindrical fruiting structures (stromata) supporting lateral rows of yellow, fruit bodies (perithecia) of *Cordyceps tuberculata*. Image: © Harry Evans.

The limited data gathered so far suggest that all invertebrate groups are commonly subject to fungal attack. For example, beneath the soil lie fungal traps – sticky knobs and networks, and constricting nooses of fungal threads – that attract and ensnare nematode worms. The unfortunate worms are then consumed by enzymes released by the pathogen. These predatory fungi also include parasites of rotifers, amoebae and other small terrestrial and aquatic animals.

Traditional mushroom hunting will, of course, reveal none of these invertebrate–fungal interactions, but microscopic examination can uncover mummified animal cadavers (nematodes, rotifers, mites) with delicate spore-producing structures emerging from them. These structures rarely exceed 50 μm in length, while those emerging from bird-eating

Mygale spiders in the Amazon, in contrast, can extend up to 50 cm. All are members of the amazingly diverse *Cordyceps* complex, which has recently been split into a number of genera following molecular studies. In pristine forests, their colourful and frequently bizarrely shaped fruit bodies advertise these fungi, thrusting through the soil and leaf litter and emerging from fallen wood. Often the host is hidden in the substratum, but a careful excavation will reveal all.

On smaller arthropods (animals with external skeletons such as insects and spiders), a diligent, 'hands-and-knees' search is necessary to uncover the myriad of *Cordyceps* associations. More than 500 species have been described thus far, but this is probably only 'the tip of the iceberg' since personal experience has shown that it is not too difficult to collect a handful of novel species in the course of a single day. Clearly, much remains to be discovered, and described. Tragically, with rapid deforestation we are running out of time to interpret these interactions within the framework of the super-organism that constitutes the tropical primary forest.

What we do know is that any factor that impacts on invertebrate survival will resonate throughout the ecosystem – no more so than in the case of slayers of colonial insects, especially ants, because of their over-riding influence on forest structure and function. Preliminary evidence from studies of fungal pathogens of ants suggests that there is a high degree of specificity and that disease pressure is continuous. Clearly, such social insects are especially vulnerable to infectious diseases, and therefore must have evolved mechanisms for reducing their impact. These include behavioural changes; for example, it has been found that infected workers in tropical forests tend to die away from the ant trails and nests: arboreal ants descend and seek hiding places whilst ground-dwellers climb. This was graphically illustrated by the discovery of *Cordyceps*-infected ants clinging to vegetation in the forest herb layer which turned out to be a new and strictly arboreal ant species of the high forest canopy.

Are these complex ant–fungus interactions only seen in tropical ecosystems? Until recently, the answer would have been yes, since temperate ecosystems have been much more intensively investigated, not least their ant ecology, and there has been no indication of fungal pathogens being implicated in ant population dynamics. However, evidence is emerging that variants of these highly specialised tropical ant pathogens also occur in the UK, only being discovered following laboratory incubation of living ants or of cadavers found in the field. It would seem that these ant–fungal associations are thus extremely ancient, pre-dating continental drift. But how do ants, and invertebrates in general, become infected?

The Chinese appear to have been the first to report on the subject of entomopathogenic (insect-killing) fungi, since *Ophiocordyceps sinensis* – a pathogen of caterpillars – has been used in traditional medicine for centuries (see Chapter 10). However, their erroneous interpretation of the association is reflected in the local name "summer plant – winter worm", from the fact that it was thought to be a single organism: the mummified larva was thought to

be the root from which the shoot (or *Cordyceps* fruit body) grew in the warmer months.

A similar assumption was made by European naturalists in the 18th century, some of whom even took this as evidence of transmutation. This was flagged by the famous Victorian mycologist Mordecai Cooke who, in his classic treatise of 1892 on entomogenous fungi (fungi that grow on insects) *Vegetable Wasps and Plant Worms*, commented on the misinterpretations of earlier collectors: "He sees in these plant-animals a proof of the passage and mutation of animal species into the vegetable, and reciprocally from the vegetable to the animal". It must be remembered, of course, that only

Left: Fungal infection can cause the behaviour of some insects to change. This onion fly, infected by *Erynia*, is showing classic 'summit disease' symptoms, whereby the host climbs the plant prior to death and remains attached by rhizoids to the tips. Rings of spore-bearing cells have burst through the abdominal sutures. Image: © Harry Evans.

Far left: Fungal infections often change ant behaviour. Arboreal ants may descend to seek hiding places and ground-dwelling species may climb. This may be the insects attempting to limit the infection spreading to other members of the colony. Here a ground-dwelling ponerine ant has climbed and has died clinging to a tropical forest shrub. Fruit bodies of the pathogenic fungus *Ophiocordyceps kniphofioides* are emerging from the neck region. Image: © Harry Evans.

Left: Fruit bodies of the fungus *Ophiocordyceps lloydii* var. *binata* are bursting from the thorax of small arboreal formicine ants attached to leaves of a tropical forest herb. Image: © Harry Evans.

Right: The European silkworm industry was ravaged by the white muscardine fungus pathogen *Beauveria bassiana*. Image: © Science Museum/SSPL.

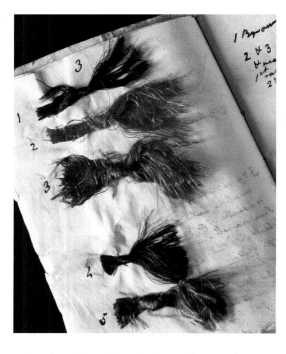

Below: A fly that has been colonised and killed by *Entomophthora muscae*. Spores (ballistospores) have been shot out from the body ready to attach to a passing fly. Image: © David Mitchell.

effective measures to control the disease, which was ravaging the European silkworm industry, but by careful experimentation he also proved that the fungus *Beauveria bassiana* was the cause and not the consequence of the disease. This pioneering work laid the foundation for the study of infectious diseases.

Hitting the target

The chances of a fungal spore reaching the target insect – which is often specific and certainly of low incidence in primary ecosystems – must be vanishingly small and remains poorly understood. What we do know about the sophistication of the mechanisms involved provides strong evidence of coevolution over a long period of time.

Fungi of the order *Entomophthorales* employ a range of spore types in order to be the successful insect-destroyers that their name suggests. The advanced heavy armoury is provided by ballistospores – spores fired explosively by a sudden pressure release – that are designed to hit the prey. Next, secondary 'insurance' spores emerge from the ballistospores when they fail to connect. These secondary spores stick to any unwary insects that tread on them – a sort of delayed action landmine. Additional spore forms also exist for dispersal (often in water) and survival. Is this a game of chance or has evolution provided underlying strategies of attack and defence?

The example of the arboreal tropical ants moving to the forest floor when infected by fungi is suggestive of a defence mechanism in the form of behaviour that would reduce the spread of infection among the ant colony. From preliminary studies, it would appear that the fungi have in a similar way evolved strategies of attack. For example, the biting simuliids (blackflies) of fast-flowing streams and rivers are hosts to species of *Erynia*; infected flies die clinging to rocks bordering these watercourses, and specialised anchoring hyphae or rhizoids emerge from the body and fix it firmly to the rock surface. Ballistospores are released from the corpses between precise hours in the late afternoon coinciding with healthy flies arriving to

a few decades earlier the 'germ theory of disease' had not been accepted and the theory of 'spontaneous generation' was still in vogue.

Proof of germ theory of disease has often been attributed to Louis Pasteur in France and Robert Koch in Germany. However, in reality, Agostino Bassi in Italy had pre-empted them by several decades. In an obscure publication dealing with the white muscardine disease of silkworms, not only did he recommend

Far left: The stinkhorn (*Phallus impudicus*) is a saprotrophic fungus that rots dead wood. The fruit body smells like rotten flesh, attracting flies that feed on it. Sticky fungus spores become attached to the flies and are spread elsewhere. Sometimes muscoid flies infected with another fungus (*Entomophthora*) die affixed to the 'head' of the stinkhorn. Image: © Patrick Hickey.

Left: Stinkhorn fruit body that has had many of the sticky brown spores removed by visiting flies. Image: © Patrick Hickey.

mate and rest. Presumably some of these are hit during the evening barrage. Casual observations have also revealed that muscoid flies infected with *Entomophthora* may die affixed to the head of the stinkhorn fungus (*Phallus impudicus*), which must provide a potent trap for incoming flies attracted by the rotting-flesh smell of the fungus.

Invertebrate killers are also found among the sac fungi – *Ascomycota*. This group of fungi seem to have developed similar weapons of destruction and similar tactics to make best use of them. For example, although many of the infected hosts of *Cordyceps* species hide in the soil, leaf litter and vegetation of the rain forest, the phototropic (growing towards light) fruit body of the fungus rises from the depths and deploys its weaponry laterally or from the tip. The ammunition storehouses or perithecia (see Chapter 1) may be immersed or on full show and the ascospores that are shot from them resemble fusillades of arrows. Is this a random event? Coevolution would suggest not. Why are the fruit bodies often brightly coloured? Do they serve to attract the host or a secondary host, or even an intermediate vector?

We do know that the ascospores of at least some species behave like the ballistospores of species of *Entomophthorales* (mentioned above), because if they fail to hit the target they can produce sticky spores on needle-like projections for a second attempt at reaching a passing host. Others may germinate to produce another form of the fungus, with spores capable of immediate dispersal, or with a separate existence on dead insect parts found in forest habitats, using this as a base for mass production of infective spores. For colonial insects, such as scale insects and whiteflies,

Below: Multiple, brightly coloured fruit bodies of an unnamed *Cordyceps* species emerging from the body and legs of a forest stick insect (Phasmidae). Image: © Harry Evans.

pathogenic *Hypocrella* species produce slime-covered spores that ooze onto the leaf surface, infecting neighbours in the colony and through rain-splash and leaf drip reaching and often destroying all colonies within that shrub or tree canopy. The ascospore projectiles serve to reach distant colonies within the forest.

Clearly, the two spore types of *Hypocrella* species and the complex array of spores produced by the *Entomophthorales* reflect the sophisticated relationships these entomopathogens have with their hosts. What we observe is the finely balanced product of an evolutionary arms-race of attack and defence. The very recent accidental finding of a rare beetle larva beneath tree bark in the UK corroborates this complexity and gives an inkling of how little we still know about these fungal life cycles. Emerging from the larva, and ultimately through a bark fissure, were two distinct spore-producing structures, one associated predominantly with arthropod hosts, the other with nematode hosts. This suggests that the fungus (originally named differently on different hosts – *Hirsutella* and *Harposporium*) can switch between dramatically different host animals.

Breaching the defences

Until recently, it was still open to debate exactly how these fungi – once reaching their hosts – infected, colonised and killed them. The commonly held belief was that spores or hyphal fragments were ingested and subsequently infected the host via the intestine. However, we now know that fungal hyphae penetrate invertebrates from the outside in a series of stepwise processes. This is unique to fungi since all other pathogens enter via the mid-gut region.

The outer covering or exoskeleton of arthropods is composed not of cells but of several layers of wax, lipoproteins and antifungal compounds, creating a hard, impervious structure which is considered to have been central to their evolutionary success. However, entomopathogenic fungi have developed sophisticated mechanisms to breach this formidable defence. Firstly, the spores land on and adhere to the inhospitable surface thanks to a combination of sticky mucilage and electrostatic forces, often forming a limpet mine-like structure on germination. From this 'beach-head' the fungus then punches through the multi-layered wall using both mechanical and chemical pressure (enzymes).

Arthropods use their immune system to fight back and have evolved a defence response to repel invaders involving specialised cells called haemocytes which ingest and destroy alien microorganisms. As an aside, the immune system of invertebrates functions in a similar way to that of vertebrates. It was earlier work with the green muscardine fungus (*Metarhizium anisopliae*) in Russia in the 1870s that prepared Eli Metchnikoff for his discovery of phagocytosis in humans and his theory of 'cellular immunity' for which he received the Nobel Prize in medicine.

Once inside the host's body cavity (haemocoel), and having by-passed the defences, colonisation begins. The two major groups of fungi – *Zygomycota* and *Ascomycota* – rely on somewhat different methods to kill their prey. The *Zygomycota* use rapid mycelial growth to claim, and ultimately slay, their host. The animal is literally choked to death by mycelium blocking the breathing holes (spiracles) on the body of most insects. Certain of these fungi manage not to trip the post-infection alarms so that the host is kept alive and functioning 'normally'. In effect, the fungus mimics the 'Alien' of cinema fame as it bursts out of the body!

Some poor unfortunates, especially cicadas, even attempt to mate – thereby spreading the disease – even though emasculated.

In the *Ascomycota*, however, colonisation of the host is more intricate. Here the fungi exist in two morphologically and physiologically different states: the budding, yeast-like form colonises the body cavity (haemocoel) and produces the toxins to kill the host, then the thread-like mycelium, and swollen hyphal bodies, induce mummification and provide the food base for spore production. In addition to producing toxic chemicals to kill the insect, the fungus also makes powerful antibiotics to prevent opportunistic microorganisms and animal scavengers from claiming the spoils once the victim is dead.

The fungus thus becomes a powerhouse of novel, highly targeted chemicals ready to launch the second wave of attack from its secured base. In many cases, particularly in *Cordyceps* species, this may be delayed until the onset of favourable environmental conditions for spore release and infection (hence the winter worm–summer plant scenario mentioned earlier). Moreover, it may be staggered over time, with sequential releases of spores: this provides an insurance policy

Far left: Tropical *Cordyceps* fruit body. The necks of the spore-producing perithecia – the ammunition stores – are visible. Image: © Harry Evans.

Left: Fruit bodies of an unnamed *Cordyceps* species emerging from the lair of a trapdoor spider host. Image: © Harry Evans.

for success, dependent on a series of carefully managed forays, rather than risking all in a single massive attack.

In agricultural (artificial) conditions, however, the strategy is different because the targets often build up to abnormally high levels – pest outbreaks – and enormous amounts of light weapons and ammunition are needed quickly. In these situations, fungi of the *Ascomycota* such as *Aschersonia*, *Beauveria*, *Lecanicillium* and *Metarhizium* seem to have opted for short-term, heavy spore production. In effect, stability has been sacrificed for short-term gain. Can these risk-taking, nomadic hordes now be organised into armies in the long-term battle against man's pests?

Below: Some fungi pathogenic on insects can be used by man to control insect pests on crop plants. Here, *Lecanicillium lecanii*, the original 'farmer's friend', is devastating colonies of the coffee green scale pest. Image: © Harry Evans.

Controlling pests

The first documented scientific attempts to use entomopathogenic fungi as weapons against arthropod pests took place during the latter half of the 19th century in Russia and the USA. The white muscardine fungus of silkworm notoriety as well as the green muscardine fungus were mass produced and the powdery spores applied as dusts against pests as diverse as sugar-beet weevils, Colorado potato beetles (*Leptinotarsa decemlineata*) and corn chinch bugs (*Blissus leucopterus*). At the beginning of the 20th century, slime spores of the brightly coloured *Aschersonia* species were sprayed against infestations of scale insects and whiteflies in the citrus orchards of Florida, where they became known as the 'friendly fungi'.

It should be remembered, of course, that these pioneering attempts at pest control pre-dated the use of the faster-acting, broader-spectrum chemical insecticides that soon replaced them, and seemingly confined them to history. In truth, the technology of mass production, formulation and application of fungi was insufficiently advanced to achieve consistent, cost-effective control, and thus to compete with chemicals.

Application of crude preparations of the most common entomopathogens continued in various countries, with mixed success – in particular in China in the 1950–70s, where unregulated cottage industries produced large quantities of *Beauveria bassiana* for use principally against rice and forest pests. Famously, or infamously for those with environmental or health and safety concerns, this 'terra-mycota' army was even deployed in mortar shells and fired over the forest canopy, in an attempt to reach defoliating insects! Brazil also tried to commercialise entomopathogenic fungi during the 1970–80s, their potential value being recognised by the local name 'O amigo do fazendeiro' (farmer's friend). Here products based mainly on *Metarhizium* species were applied against insect pests of pasture and sugar cane. Interest has waxed and waned since, and although minor battles were won, mycoinsecticides never became the weapon of choice in the war against insect pests. Similarly, efforts to use specialist pathogens of plant parasitic nematodes, mainly species of the order *Hypocreales*, have been on-going since the latter half of the last century, but have thus far not achieved front-line status.

The world has now moved on and many of the tried and tested insecticides, acaricides, nematicides and soil fumigants have lost favour, due to increasing awareness of the negative impact of these chemicals on the environment and the fact that pests become resistant to chemical pesticides. The 'friendly fungi' are once again receiving attention and are being resurrected to fill the void left as more pesticides fail to meet regulatory standards.

Already there have been some outstanding successes, most notably 'Green Muscle®' – based on *Metarhizium anisopliae* dispersed in an oil – which, although not as fast acting as chemical competitors, can compete in terms of efficacy. Improved technology of formulation and ultra low volume application means that this mycoinsecticide can be used against locusts and grasshoppers – the bane of subsistence farmers throughout Africa – in the driest of habitats.

Although considerable scientific hurdles have been overcome, the political and logistical ones remain: who will manufacture, store, distribute and apply a product against a periodic pest over large areas of land of low productivity and which will benefit only resource-poor farmers? A similar dilemma faces those attempting to develop fungal products against mosquitoes, as well as other fly vectors of human diseases. Many fungal groups responsible for major diseases in mosquito populations worldwide have been identified: *Chytridiomycota* (*Coelomomyces* species),

Zygomycota (*Conidiobolus*, *Entomophthora*, *Erynia*, *Smittium* species), and *Ascomycota* (*Beauveria*, *Culicinomyces*, *Metarhizium*, *Tolypocladium* species). Some of these show highly desirable and unique characteristics compared with chemicals, including being specific to their hosts, being actively dispersed by adult females to breeding sites, and having residual activity and persistence without environmental harm.

Clearly, however, significant investment in new technology is needed, as well as political will, to overcome the previously identified constraints to using living organisms as biopesticides. In the future these potential fungal saviours could be added to the now limited arsenal of weapons which can be deployed against pests, and thereby make a significant contribution to the management of some of the most intractable of human diseases, including dengue, filariasis and, above all, malaria.

The secret may lie in trying to fool the insects into not recognising their fungal foes until it is too late, perhaps by disguising them in order to gain entrance into the vulnerable nest, or by employing a Trojan-horse tactic to breach the defences. To achieve victory, the battle plans will need to be both flexible and innovative. And, as with all potential mycopesticides, the weaponry will need upgrading; perhaps new formulations and application techniques could

Above: The green muscardine fungus (*Metarhizium anisopliae*) has been used with outstanding success as a mycoinsecticide. Here it is seen emerging from the body sutures of a chrysomelid beetle. Image: © Harry Evans.

Left: Fruit bodies of *Ophiocordyceps amazonica* with strawberry-like heads emerging from a forest grasshopper. Image: © Harry Evans.

Above: The chytrid fungus *Batrachochytrium dendrobatidis* (*Bd* for short) is causing a devastating disease of amphibians in many countries. Three species of frog have already become extinct as a result of this pathogen. Here, midwife (*Alytes obstetricans*) and common toads (*Bufo bufo*) have been killed by *Bd*. Image: © Matthew Fisher.

help convert the 'ascospore arrows', 'conidial grapeshot' and 'ballistospores' into armour-piercing missiles with faster and more reliable kill. Mycelial 'cluster bombs' with the ability to rapidly produce large quantities of spores is another option, along with impregnating plants with cryptic (endophytic; see Chapter 3) mycelium to deter insect herbivores. It is no coincidence that many endophytic and entomopathogenic fungi share a common evolutionary pathway.

Whatever the future holds for mycopesticides, it is in our long-term interest to invest more resources in investigating these animal–fungal associations. This could lead not only to a greater understanding of ecosystem functioning but also, more pragmatically, to the discovery of novel chemicals which help these fungi to maintain their unique lifestyles and which could be of immense benefit to both medicine and agriculture.

Vertebrate slayers

Some fungi are also able to grow extensively in vertebrates and even to kill them. Perhaps the most devastating example is the chytrid *Batrachochytrium dendrobatidis* (called *Bd* for short). *Bd* causes a disease of amphibians which has resulted in extinction of at least three species – the Panamanian golden frog (*Atelopus zeteki*), the Australian gastric brooding frogs (*Rheobatrachus* species) and the sharp-snouted day frog (*Taudactylus acutirostris*) – and is suspected to have caused extinctions of many

more. This new disease is probably spread by the global trade in amphibians, and made worse by climate change. The ability of *Bd* to infect and cause disease in amphibians depends on temperature and moisture availability. Climate change may have opened an 'environmental window' that has allowed this pathogen to flourish, so this devastating disease may indirectly be a consequence of man's activities.

In humans, there are half a dozen or so extremely nasty fungal diseases, with equally horrible sounding names, including aspergillosis, candidiasis, histoplasmosis, coccidiomycosis and blastomycosis. Of more than 200 species of *Aspergillus* probably fewer than 20 cause disease in animals, including humans. They can be thought of as opportunists, growing inside people if they happen to find themselves there but well able to live in the natural environment.

If people are exposed to abnormally high spore loads, often through working in confined spaces such as areas for storing agricultural produce, they may eventually succumb to bronchial diseases caused by *Aspergillus* species – notably farmer's lung, a debilitating condition caused by fungi that grow happily at the elevated temperatures found in the human body. New harvesting methods and improved storage protocols seem to have reduced the incidence of this disease by reducing the amount of fungal inoculum, especially of the ubiquitous *Aspergillus fumigatus*, in the agricultural work environment. However, this species in particular is currently a major cause for concern because of its lethal invasion of immuno-compromised patients, such as those suffering from AIDS. It should be stressed, however, that all these are artificial situations created by man's recent and continuing social and economic development. Nonetheless, annually *Aspergillus* infections cause around 700,000 deaths worldwide and *Cryptococcus* infections around 400,000.

Candida albicans is present in our moist bodily orifices all of the time and mostly we do not even notice it, as our immune system keeps it under control. In some circumstances its yeast cells can proliferate and cause 'thrush'. Typical symptoms include white 'furry' patches and the skin becoming inflamed and uncomfortable. However, most cases are fairly easily treated with an appropriate cream or gel. *Candida* tends to invade the body causing serious problems only in individuals whose immune system is severely debilitated.

While infections caused by *Aspergillus* and *Candida* species occur all over the world, some fungal diseases seem to be more or less confined to certain geographical areas. Histoplasmosis is found in the Mississippi and Ohio valleys. The fungus that causes the disease – *Histoplasma capsulatum* – grows well in wild bird droppings, chicken manure and bat guano, all of which are rich in nitrogen. Spores (conidia) are inhaled and grow in a yeast form in the lungs. Symptoms are flu-like, but occasionally the infection can go on to produce a progressive lung disease rather like tuberculosis. In 95% of cases no obvious symptoms are produced, and millions of people will have had infections without even knowing it. Coccidiomycosis is a fungal disease found in the southwest USA, Central America and northern South America. The fungus responsible – *Coccidioides immitis* – thrives in dry, salty soils typical of desert areas (though strangely it is not a problem in deserts of Africa or Asia). The clinical symptoms are similar to histoplasmosis, and once again millions of people have contracted the illness and recovered without treatment, though again there are occasional fatalities.

Blastomycosis, caused by *Blastomyces dermatitidis*, is found in the eastern USA and Canada. The fungus grows in rotting plant material and infection occurs when the spores become airborne. The spores again infect and

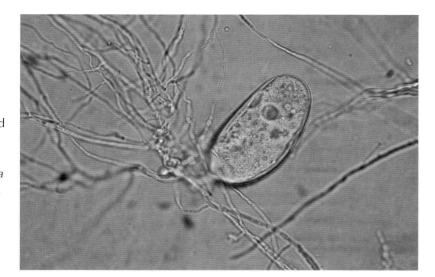

multiply in the lungs. From there the fungus can spread to other organs via blood and lymph. The disease can be acute, resembling pneumonia, chronic tuberculosis or lung cancer, and can result in acute respiratory distress. It can be fatal, especially in immuno-compromised patients. However, even in areas where the disease is prevalent only one or two people in 100,000 get it.

Socialists

The story we have told so far gives the impression that fungi are only involved in harmful relationships with animals, but this is far from true. Fungal mycelium and fruit bodies are very nutritious, not just for humans (see Chapters 8 and 9) but also for other animals. For example, in Australia truffles are an important food source for a range of mammals.

Above: A chytrid that inhabits the gut of ruminants. These fungi aid in the digestion of the vegetation eaten by the animal. Notice the hyphae that extend from the main body of the fungus. These grow into and break down ingested vegetation. Image: © Gareth Griffith.

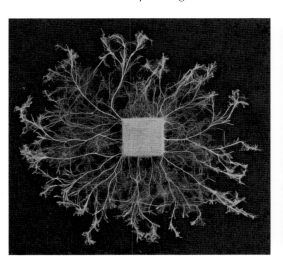

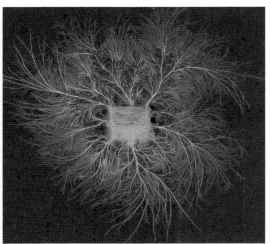

Far left: Growth of fungal mycelia is influenced by invertebrate grazing. This springtail-grazed mycelium is noticeably different in appearance to the ungrazed example to its right. Image: © George Tordoff.

Left: Ungrazed fungal mycelium showing different growth patterns to the grazed example to its left. Image: © Alaa Alawi.

Above: Many species of termites cultivate fungi in their nests. This is of mutual benefit to both the fungus and the termite. The termite brings dead plant material to the nest, and the fungus breaks it down and feeds on it. The termites then feed on part of the fungus. Shown here is the fungus comb with highly nutritious white mycotêtes produced by the fungus and fed on by the termites.
Image: © Karen Machielsen.

In soil habitats everywhere many invertebrates use fungi as food, especially arthropods but also slugs and snails, earthworms and springtails. These organisms feed directly on mycelium, or indirectly when they ingest rotting plant material, often with dramatic effects on the fungus.

Fruit bodies provide a food source and breeding grounds for a wide range of invertebrates. The smell of fruit bodies varies depending on the species, and invertebrates often use this to home in on their favourite fungus. The stinkhorn (*Phallus impudicus*) smells of rotting flesh and is attractive to flies. The flies feed on the fungus, and as they do so the sticky spores attach to their legs and body and are then spread far and wide.

Over millions of years fungi have formed innumerable liaisons with invertebrates in which both partners benefit. For example, 'ambrosia' fungi (lots of different ascomycete species) are carried to trees in special organs in the bodies of ambrosia beetles (belonging to several different groups). The fungi line the galleries formed in wood by the burrowing beetle larvae, and are grazed on by the larvae. Some invertebrates

carry disease-causing fungi to their hosts; for example, *Scolytus* beetles carry *Ophiostoma novo-ulmi* to elms, where the fungus causes Dutch elm disease. However, those that have the greatest overall ecological impact are the interactions with social insects, particularly ants and termites.

In tropical regions at least, mutually beneficial relationships between invertebrates and fungi may be the principal drivers of ecosystem functioning. For example, the leaf-cutting ants of the Neotropics are considered as keystone species since they are the most important primary consumers in terrestrial ecosystems, especially forests and grasslands. Pivotal to their success is the association with fungi such as *Leucoagaricus* species. The ant supplies leafy compost for fungal growth and in return the fungus produces 'goody bags' of swollen, nutrient-filled hyphae for ant consumption. Indeed, these ants are consummate mycologists, tending their fungal gardens to keep them free of invasive microbes.

Similarly, termites provide wood to their fungal associates in the nest; for example, *Macrotermes* species tend their gardens of

Termitomyces species and reap the fungal harvest from this mutualism. The termites do not have the enzymes necessary to decompose woody materials but the fungi do (see Chapter 2). Again they convert low nutrient plant material into highly nutritious, nitrogen-rich fungal mycelium. In fact, they produce special structures – mycotêtes (literally fungus head) – that are eaten by the termites.

These important animal engineers of natural ecosystems could not function without their mycological socialist partners (or symbionts) since the fungi are the real, but hidden, power behind the queen's throne, supplying the energy that drives the colony. Termites and leaf-cutting ants, when in contact with man and his ecosystems, can cause considerable conflict of interest by destroying his crops and infrastructure. Ironically, mycoinsecticides are now being targeted at both ant and termite pests worldwide.

The relationships which fungi have developed with animals, including man, are diverse. They are often the product of coevolution over millions of years. The attack and digestion of invertebrates, sometimes alive, is an extraordinary example of reality being at least as strange as the horror scenarios of science fiction (see Chapter 7). Some of these interactions may seem to be of largely academic interest. However, there is no telling where the next wonder drug, for example, may come from. Based on past experience, the potential for future human uses of fungi is enormous. To allow this potential to be explored and unlocked we need to preserve areas where the delicate balance of nature, with all its interconnections, can continue to be studied (see Chapter 10).

Let us finish with an example of a fungal mutualism that is of great importance to man. The chytrid fungi live in the guts of ruminant herbivores such as cows and sheep. These fungi, along with bacteria and protozoa, are crucial to the digestion of the vegetation these animals consume. As in the case of ant and termite mutualisms, the ruminants provide the fungus with a supply of food and the fungi break down plant fragments in the animals' stomach, allowing them to thrive on the plants they eat. Without fungi we would have no meat, no milk, no wool, no leather, and no herbivores!

Ophiocordyceps cantharelloides species profile

Thirty years ago while dealing with fungal pathogens on a tea estate in Amazonia I accepted an invitation to camp in the densely forested scarp towering above the estate. It was the first time I had camped out in tropical forest and nearly the last, as the stillness was shattered by a massive tree falling metres from the tent. But any thoughts about the near miss were quickly dispelled by the richness of the forest fungi, especially those I had a special interest in – the *Cordyceps* complex. The star of the show, which I have since sought for and not found, was a strange 'mushroom' seemingly growing from the smooth bark of a young tree at about head height. Impulsively, I removed it to have a closer look with a hand lens. I could see the characteristic fruiting structures of an ascomycete, so possibly an example of the *Cordyceps* fungi that parasitise insects and other invertebrates. On closer inspection, the stalk was emerging from the wood via a perfectly round hole. Could there be an insect host within the tree? I duly began the excavation, but the wood was iron hard and it took several hours of clumsy machete wielding to follow the stalk and reveal the pencil-sized beetle larva deep within the living tree. I had my specimen, or at least the several pieces into which it had been butchered. Eventually, it was named *Cordyceps cantharelloides* in view of its amazing likeness to the chanterelle mushroom, and is now lodged in the basement of the Kew Herbarium. Since this submontane forest is one of the most threatened habitats on Earth, and the beetle host and fungus are undoubtedly rare within the system, it seems probable that this species will never be found again: a sad memorial to our accelerating forest destruction over the past four decades.

Harry Evans

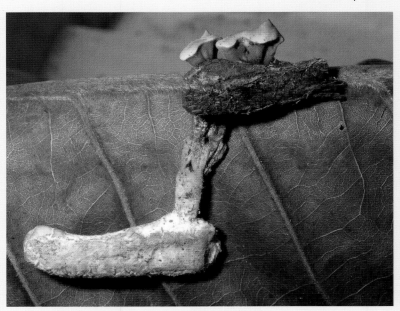

This fungus is smaller than, but similar to, *O. cantharelloides*. It has colonised and then fruited on a wood-boring beetle larva. Image: © Harry Evans.

Main photos, clockwise from this page:

Coprinus disseminatus;
Lactarius torminosus;
Xylaria longipes.
Images: © Chris Jeffree.

Facing page inset photos:
Fungus dyes (left); fungus derived medicines (right).
Images: © RBGE/Max Coleman.

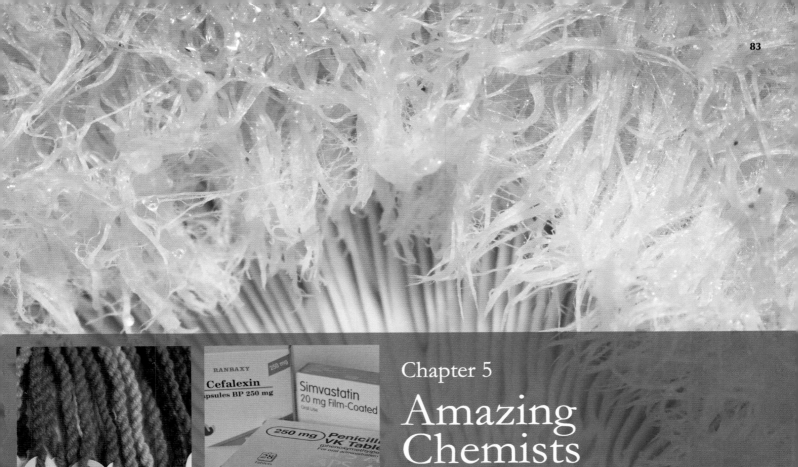

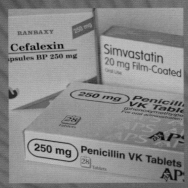

Chapter 5
Amazing Chemists

Milton Wainwright

Amazing Chemists

Milton Wainwright

Fungi often get a bad press, largely because they are seen solely as organisms that destroy our crops and stored food and, in the case of dry rot, building timbers (see Chapter 2). In addition, they also cause human disease, some of which can be life-threatening (see Chapter 4). While there is no doubt that fungi can cause considerable economic damage, they also contribute positively to the human economy.

As we shall see, fungi play an amazingly important role as 'chemical factories' which are used to produce a range of commercial products, from life-saving drugs to the more mundane food staples bread and beer. Fungi can also be consumed directly as mushrooms or as a much-modified protein source called mycoprotein, a manufactured food, widely consumed as Quorn®. In addition, the ability of fungi to produce a wide range of substances when grown in bulk culture is exploited to provide a vast array of useful chemicals, from enzymes and fats used in the food industry, to drugs used in the treatment of cancer and in preventing the rejection of transplanted organs, to the enzyme tannase used in leather tanning. Of particular contemporary interest is the ability of fungi to produce biofuels, such as ethanol and biodiesel, which may one day replace fossil fuels.

Right: Some fungi and lichens can be used to dye yarn and material. Image: © RBGE/Max Coleman.

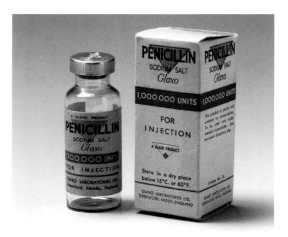

Far left: Penicillin is one of the most well-known antibiotics. Its discoverer, Alexander Fleming, won the Nobel Prize for the breakthrough in the treatment of infection. Image: © Science Museum/ SSPL.

Left: Fleming's Nobel Prize medal given for the discovery of penicillin. Image: © Science Museum/ SSPL.

Fungi in medicine

Penicillin, the most famous and important of all the antibiotics, is produced by a mundane mould called *Penicillium chrysogenum*, the kind of green growth that you might see covering jam or a slice of stale bread. The discovery of penicillin, by Alexander Fleming, and its development for medicine by Howard Florey, Ernst Chain and Norman Heatley, is the stuff of legend. When penicillin first appeared in the early 1940s it was seen as a miraculous wonder drug and it soon revolutionised medicine. Previously fatal diseases (such as septicaemia) and sexually transmitted bacterial infections (such as syphilis and gonorrhoea) were soon defeated by this metabolic product of the lowly filamentous fungus. By controlling bacterial infections penicillin allowed major developments in surgery to take place. Other fungal antibiotics, such as cephalosporin, soon followed and eventually a cure was found for the once dreaded respiratory infection, tuberculosis. The antibiotic involved here, streptomycin, was produced by a filamentous organism called *Streptomyces* which was for a long time considered to be a fungus, but is now firmly placed among the bacteria; it is from this class of organism, rather than the fungi, that most of the more recent antibiotics have been derived. While the first penicillins were natural products produced by a fungus, semi-synthetic penicillins, which possessed a wider range of antibiotic activity, were soon developed; the production of such wide-spectrum antibiotics still, however, requires the product of industrial-scale fermentation of the fungus *Penicillium*.

Cephalosporin, the other major fungal-produced antibiotic, has an interesting history, being obtained from a filamentous fungus originally called *Cephalosporium* which was isolated from a sewage outfall in Sardinia. This provides a good example of how fungi producing important metabolites can be isolated from the most unlikely of places; it is not surprising then that chemical–pharmaceutical companies scour the Earth's unusual environments for novel fungi which they hope will produce new commercially useful chemicals. A few other fungal antibiotics have been produced, such as griseofulvin which is used in the treatment of fungal diseases like athlete's foot, but nothing compares in importance to penicillin, an antibiotic which is still widely used in medicine despite the fact that many pathogenic bacteria (most notably MRSA – methicillin-resistant *Staphylococcus aureus*) have become resistant to it.

Below: Fungi produce several hugely important medicines including the antibiotics penicillin and cephalosporin. They also produce statins, used to reduce cholesterol. Image: © RBGE/Max Coleman.

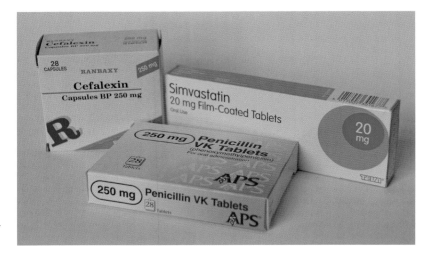

A fungal product is even being used in the treatment of breast cancer. Called taxol, it was originally isolated from yew (*Taxus*) trees, but a number of endophytic fungi (i.e. those growing inside plants; see Chapter 3) have been shown to produce this anti-cancer compound, potentially the most important being *Pestalotiopsis microspora*. Taxol is also antifungal and it may be that yew trees provide a home for an endophytic fungus that produces taxol in order to prevent pathogenic fungi from infecting the trees.

Many people across the world currently use statins to lower cholesterol levels and so reduce the risk of heart disease. These compounds (such as lovastatin) are produced by the filamentous fungi *Aspergillus terreus* and *Penicillium citrinum*. They inhibit the enzymes needed to make cholesterol and are credited with saving 700 lives a year in the UK alone.

Another fungal compound, called cyclosporin, is widely used as a drug to help prevent the rejection of transplanted organs, such as the heart and kidney. Cyclosporin is produced by yet another filamentous fungus, *Tolypocladium inflatum*. The drug works by preventing the production of lymphocytes (white blood cells), which are an important part of the immune response which enables the body to

detect and deal with infection. Without this drug the donor organ is often recognised as foreign and destroyed by the patient's immune system. A similar immunosuppressant compound, called gliotoxin, is produced by the filamentous fungus *Aspergillus fumigatus* which, incidentally, also causes farmer's lung disease (see Chapter 4). Gliotoxin was originally noticed for its antibiotic activity, but was soon found to have other medically useful properties. One wonders what other useful medical compounds have additional properties that have yet to be discovered.

Although cancer is far from defeated, there have been considerable improvements in the treatment of this dreaded disease. Anti-tumour drugs, including lentinan, are produced by a fungus which is widely eaten in Japan, *Lentinula edodes*. Lentinan is used in the treatment of stomach cancers and is also used as an antihistamine in the treatment of hay fever. Mention should also be made of ergot, a chemical produced by several fungi, most notably *Claviceps purpurea* when growing on rye. Ergot has long been used in traditional medicine to aid abortion and prevent bleeding in childbirth; ergot-derived drugs, including pergolide, have also been used more recently in the treatment of Parkinson's disease, although their use is restricted by their severe side effects.

Finally, fungi can also be used to transform steroids into novel variants in order to make them more active in the treatment of diseases like rheumatoid arthritis.

Fungi in the food industry

The most obvious use of fungi as a food is the consumption of mushrooms, mainly the white button mushroom (*Agaricus bisporus*). However, our tastes are changing and some 20 different types of mushroom are now produced on an industrial scale, including ones once considered to be somewhat exotic, such as the shiitake (*Lentinula edodes*). The worldwide market for mushrooms was estimated in 2001 at some US$40 billion, while in the UK the annual market is around £320 million.

A less obvious use of fungi in food production is in the fermentation of beer, wine and other alcoholic drinks, which relies upon the ability of the yeast *Saccharomyces cerevisiae* to ferment sugars to ethanol. The traditional fermented alcoholic beverage of Japan – sake – is also produced by a fungus, in this case not a yeast but the mycelial fungus *Aspergillus oryzae* grown on polished, steamed rice.

A by-product of the brewing industry is food yeast. Rich in vitamin B, this is marketed in the UK as the popular sandwich spread Marmite® and elsewhere as a similar product, Vegemite®. Yeasts are also used in the baking industry, due to their ability to produce carbon dioxide from sugars, which causes bread and other baked products to rise.

Filamentous fungi are also used to add flavour to cheeses, producing the classic blue-veined varieties such as Blue Stilton®, Roquefort and Danish Blue; the fungus involved is *Penicillium roqueforti*. Such blue cheeses have a long history and were described as being delicious by the Roman naturalist and philosopher Pliny, as early as AD 79. The texture of Camembert is also produced by a fungus. In fact, fungi are now used in the manufacture of most cheeses these days; while in the past rennet was used to coagulate the milk, nowadays the fungally produced chimosin is employed.

Wine (and other alcoholic drinks), bread and cheese are all made with the help of fungi. Natural yeasts on the skins of grapes are often used in wine production.
Image: © RBGE/Lynsey Wilson.

It may come as a surprise to many people that soy sauce is another fungal product. In this case the fungus *Aspergillus oryzae* and the yeast *Pediococcus* are used to ferment soya beans. Other fungal foods which appear exotic to Western eyes include Indonesian *tempeh kederle*, a white mould covered cake made from soya shoots, *idii paty, lao chao, ang-kak* and *oji*, a porridge produced in West Africa from fermentation of corn by bacteria, yeasts and filamentous fungi.

Protein-rich animal feeds can also be produced by fungi. Fungi such as *Trichoderma* species can be grown on all kinds of organic wastes, and these are fed to animals to provide a food rich in protein and carbohydrates. Fungi can also be grown in this way on food grains like barley and thereby improve the protein content. Perhaps the most intriguing example is provided by the so-called Pekilo process, whereby the filamentous fungus *Paecilomyces variotii* is grown in Finland on spent sulphite liquor, a waste product of the paper industry. In addition to producing valuable protein, the process also reduces pollution from the paper mills. Protein can also be produced by growing fungi on whey, cassava and waste oils from refineries.

Mycoprotein (marketed as Quorn®), a popular food of vegetarians, is also produced by a filamentous fungus, *Fusarium venenatum*. Although advertised as a relative of the mushroom family it is an ascomycete not a basidiomycete (see Chapter 1). Quorn® was developed in the late 1960s and has since become a popular alternative to meat; it gets its meat-like texture from the filamentous nature of the fungus.

Fungi are also widely used to produce chemicals for the food industry, most notably citric acid, a common component of foods, particularly fizzy drinks. Citric acid, produced by the filamentous fungus *Aspergillus niger* (the world production now mostly comes from China), is also used in the manufacture of detergents.

Lipids or fats are also produced by yeasts. A species of *Candida*, for example, can produce a replacement for the relatively expensive cocoa butter used in the manufacture of chocolate. Filamentous fungi also provide alternative sources of the active ingredient in the dietary supplements primrose oil and fish oils.

Recent concern about the use of artificially produced food colourants has led consumers to demand more 'natural' alternatives. An example is the red colour produced by the

fungus *Phaffia*, which can be used to impart a desirable red colour to trout. In the Orient, a fungus called *Monascus* has been used traditionally to produce red rice wine, and the yeast *Candida utilis* to concentrate the red beet colour from beet juice. Some mutants of *Phycomyces* produce beta carotene, while the filamentous fungus *Aspergillus niger* produces an enzyme called naringinase which can be used to remove the bitter taste from grapefruit.

Above: *Phycomyces blakesleeanus.* Some mutants of *Phycomyces* are used to produce the natural food colourant beta carotene. Image: © Patrick Hickey.

Fungi and fuel

The pressing need to move away from a reliance on fossil fuels, such as coal, oil and gas, has stimulated interest in the biofuel industry. Alcohol (ethanol), produced by yeast from sugars during fermentation, provides the most common biofuel, used worldwide but most notably in Brazil. Yeasts can ferment all kinds of sugar-rich substrates whether derived from wheat, maize, sugar beet and sugar cane, or molasses; in fact any substance from which alcoholic beverages can be made. The ethanol produced can be used directly in car engines or mixed with petrol in any combination. Most run on 15% bioethanol–petrol mixes. Ethanol has a higher octane rating than petrol and is therefore more efficient, as well as being less polluting, hence the widely used term 'green fuels'. Unfortunately, ethanol is corrosive to fuel systems, rubber hoses, etc. and is

Left: A red food colourant often used in trout is obtained from the fungus *Phaffia*. Image: © RBGE/Max Coleman.

hygroscopic; this corrosive tendency increases costs of transport and general use. Ethanol can also be used to heat homes, in flueless burners, although the heat output is somewhat lower than natural gas and electricity. Unfortunately, when all the production costs are taken into account, bioethanol currently does not produce major benefits in lowering net energy and, as yet, does little to reduce fossil fuel dependency.

An interesting new development is the production of fungal biodiesel. Biodiesel can be made from algae, but recently the filamentous fungus *Gliocladium roseum* has been found to produce the fuel when grown on cellulose wastes.

Other uses of fungi

Anyone walking through a woodland soon becomes aware of the ability of some fungi to decompose wood (see Chapter 2). They can do this because they are able to produce enzymes which decompose lignins, the very complex molecules that make wood so durable. Some basidiomycetes and a few ascomycetes are the only organisms on the planet that can break down lignin. This ability to alter very complex molecules is being used in so-called

bioremediation – the breakdown of potentially dangerous pollutants in the environment. Fungi that can decompose lignin can often also decompose complex organic compounds such as pesticides and hydrocarbons that can act as environmental pollutants.

The ability of fungi to produce these enzymes is also useful in the paper industry for what is called biological pulping. The mechanical pulping of wood to make paper is extremely energy intensive. The use of fungal enzymes to soften wood before pulping requires far less energy and has tremendous potential for reducing the energy bills of paper mills. An additional benefit is that lignase-producing fungi can also be used to break down the highly damaging pollutants produced during the paper-making process.

Fungi can also be manipulated to produce useful chemicals that they would not otherwise be able to. Genetic modification (GM) allows the genes responsible for the manufacture of a compound to be moved from one organism to another. Although this technology has met with much public suspicion, in the context of producing medically useful compounds it offers potentially enormous human benefit. The yeast used in bread-making, *Saccharomyces cerevisiae*, is one of the main species used in the production of so-called recombinant enzymes (proteins manufactured by the yeast using the genetic code of another organism). As this yeast is a food organism and considered extremely safe it is an ideal host for the production of commercial proteins. It can secrete proteins into the growth medium in large quantities and since we know a lot about manipulating its genetics it can be readily used to produce novel compounds. For example, we can clone mammalian genes in this yeast, including those that will make the yeast produce human interferon, human epidermal growth factor and human haemoglobin. The most important use of this technology to date, however, has been the production of the first safe hepatitis vaccine.

When in 1976 Geoff Pugh (a former president of the British Mycological Society, and my postgraduate supervisor) was asked to provide an inaugural address on his appointment to a professorship at Aston University, he chose the title 'Have You Thanked a Fungus Today?' In his lecture, he outlined just how important fungi are in our everyday lives. More than 30 years have passed since then and the importance of fungi in human affairs has increased beyond what anyone at that time could have imagined. True, in the 1970s we knew that fungi were important producers of antibiotics, and of course the use of yeasts to produce bread and alcoholic drinks had been known since antiquity. But the revolution in molecular genetics had not yet begun and the idea that fungi might, for example, one day produce vaccines, and biofuels that would fuel vehicles across the world, was barely considered. I hope that this chapter has convinced the reader of the importance of fungi to our daily lives, in the food we eat, in our medicines and, increasingly, in the production of low polluting alternative fuels. Far from being just agents of decay, fungi continue to make an extremely positive contribution to the human economy; we do indeed have a lot to thank the fungi for every day.

Phellinus ferreus (cinnamon porecrust) species profile

During the 1990s a well-known supermarket was concerned about ill-odours emanating from foodstuffs kept on its shelves. This led to an examination by a contact of mine of the gases emitted from not only food but also a miscellaneous range of microorganisms for which cultures were easily available. Amongst those tested was a small group which gave off measurable quantities of chlorinated hydrocarbons. When I scanned the results in the journal *Nature* what caught my eye was the fact that all those giving off these gases were members of the same group of fungi – the *Hymenochaetales*. This led me to believe that this group might possess some important properties. So I embarked on a collaborative project.

First, in a blind test, unlabelled brown paper bags containing known fungi were sent for analysis. As suspected, all the members of the *Hymenochaetales* came back positive but not a tweet from the others! We went on to look at timbers from all over the world, trying to ascertain where the chlorine came from and the distribution in the world's forests of members of this distinct fungus group. This was very exciting as it gave quite unexpected results and led us to look at the total global production by these fungi of chlorinated hydrocarbons (which are important greenhouse gases and also deplete the ozone layer). Our guestimate was 16,000 metric tons – which exceeds the deleterious gases from the world's refrigerators and freezers. Researchers in Brazil are trying to measure the gaseous flux in the field to confirm our results.

Thus felling of forest and burning the residue has several important consequences in addition to loss of trees and a reduction in the capacity of the Earth's lungs. It also produces particulate matter (linked to bronchial problems) and sulphur dioxide and carbon dioxide, and in addition – from the woody trash left behind, home to many *Hymenochaetales* – chlorinated hydrocarbons.

What a way to finish one's professional career! It goes to show that classical taxonomy should not be assigned to the dust heap as here is an example of the way it can be joined up with cutting edge science!

Members of this group of fungi are all around us; the photograph shows the cinnamon porecrust (*Phellinus ferreus*).

Roy Watling

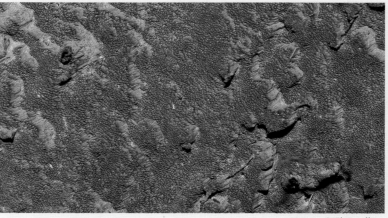

Image: © Chris Jeffree.

Main photos, clockwise from this page:

Lycoperdon species.
Image: © RBGE/Robert Unwin.

Daedalea quercina;
Ascocoryne sarcoides.
Images: © Chris Jeffree.

Facing page inset photos:

Liberty cap
(*Psilocybe semilanceata*) (left).
Image: © Patrick Hickey.

Chanterelle
(*Cantharellus cibarius*) (right).
Image: © Chris Jeffree.

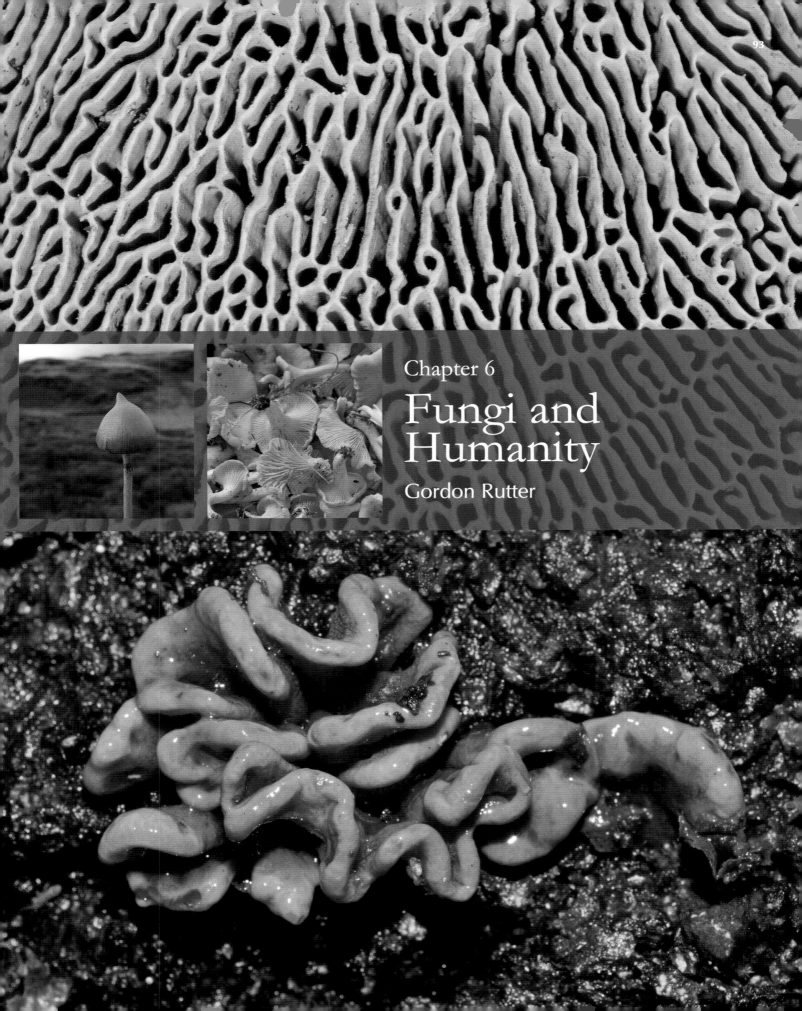

Chapter 6

Fungi and
Humanity

Gordon Rutter

Fungi and Humanity

Gordon Rutter

Right: Fungi are held in high esteem in many areas of continental Europe, to the extent that some bars have mushrooms on their advertising signs, as in this small Andalucian town of Aracena. Image: © Pat Leonard.

Below: Some fungi fruit in circles or arcs, or cause a ring of dead grass, or a ring of lush grass known as a 'fairy ring', as here with *Marasmius oreades*, in Minnesota, USA. Image: © Wally Eberhart/ Visuals Unlimited/Corbis.

Humanity and fungi have had a long relationship. In addition to the obvious benefits, such as bread and alcohol, there have also been many other important interactions. Indeed, some have gone so far as to suggest that the advent of religion itself was the product of fungi. Over the next few pages we will look at some of the many ways in which people have interacted with fungi. However, just as a mushroom is the tiniest part of the fungus appearing above ground, this short introduction only scratches the surface of this fascinating subtopic of mycology.

Fairy rings

One rich vein of humanity's take on fungi can be found in the wealth of folklore surrounding them. An obvious example is the fairy ring. Fairy rings comprise a ring of luxuriantly growing grass, or dead grass and bare earth or all of these. The ring itself is often visible for most of the year, but the fungi responsible only produce fruit bodies for short periods of time. Around 60 different species of fungi regularly produce fairy rings. The ring structure comes about as a by-product of the radial expansion typical of many fungi. A spore lands in a suitable habitat, it germinates, and the mycelium grows in a radial pattern extending at the hyphal tips forming an ever widening circle. When the conditions are appropriate fruit bodies are produced, often at characteristic times of year (St George's mushroom typically fruits around St George's Day).

In folklore the commonest explanation for the production of these bare patches of ground is – of course – fairies. After a night of merrymaking they would gather for one huge final dance – a circular dance – and such was

the ferocity of their dancing that they would scar the earth beneath them. In Shakespeare's *The Tempest*, Prospero (Act iv, scene 1) makes reference to this belief:

You demi-puppets that
By moonshine do the green sour ringlets make,
Whereof ewe not bites; and you, whose pastime
Is to make midnight mushrooms…

In fact, for proof that this is what happens we don't have to rely on the writings of Shakespeare, for in 1663 John Aubrey describes an eye-witness account of such merrymaking. Aubrey's local curate Mr Harte was out for a walk one evening when he came across a group of small individuals dancing. As he watched them dance round and round, some of them saw him; they came across and, swarming around the curate, caused him to fall over. The little people then began pinching him and humming. When they left him alone at daybreak the curate found he was inside a fairy ring.

But fairies aren't the only beings blamed – for example, in the Austrian Tyrol dragons flying in small circles let their tails drag on the ground, thereby burning a ring pattern. Or hedgehogs run round and round in circles wearing away the grass. Or moles do the same underground and their droppings fertilise the earth causing luxuriant grass growth. Or cattle crowd round a bale of straw, feeding at one end and dropping fertiliser at the other – naturally in a circular pattern, as this will allow the maximum number of cattle to get to the straw. The list goes on. In fact, the fairy rings were the crop circles of their day!

In reality the luxuriant growth is the result of the external digestion by the fungal mycelium: enzymes are released which break the fungal food into its component nutrients (see the Introduction). But fungi are messy eaters and some of the nutrients remain in the soil, acting as fertiliser for grass and so causing the characteristic luxuriant growth. In the bare-earth fairy rings the mycelia are so densely packed that it is difficult for water to penetrate, so there is not much water available for plant growth; and there is anyway very little room for plant material due to the denseness of the mycelia.

Hallucinogenic fungi

Some fungi are employed for their hallucinogenic properties. Much of the modern pioneering work on ethnomycology and the use of hallucinogenic fungi in rituals was carried out in the late 1950s by an American investment banker, R. Gordon Wasson, along with mycologist Roger Heim.

Hallucinogenic fungi are believed to be represented in Stone Age art from an area that is now the Sahara Desert but 7,000–9,000 years ago, when the images were produced, was covered with vegetation. In one series of images a line of figures, all apparently wearing hats, are holding what appear to be mushrooms. Other images from the area show masked (transformed?) figures holding fungi and apparently covered in mushrooms. In some of the cave art, rays appear to emanate from the mushrooms out to the heads of the dancers; perhaps an indication of the mind-altering effects of the fungi? Many other similar images in the area point to the importance attached to mushrooms by the native peoples.

Above: R. Gordon Wasson carried out much original research into the traditional uses of fungi by the Amerindians during the 1950s. Wasson was interested in hallucinogenic fungi and essentially gave birth to the 'magic mushroom' movement in his article 'Seeking the Magic Mushroom' for *Life* magazine in 1957. Here he is depicted in Mexico in 1955 with Maria Sabina a Mazatec curandera, or healer, who introduced him to psilocybin mushrooms. Image: Courtesy of The Tina & R. Gordon Wasson Ethnomycological Collection Archives.

Above: Liberty cap (*Psilocybe semilanceata*) is a hallucinogenic fungus whose use is illegal in the UK, where it is regarded as a Class A drug, putting it on a par with heroin. Image: © Patrick Hickey.

Right: Fungus stones are found throughout northern Central America, especially Guatemala. Their varied design has been used to enquire into the part played by hallucinogenic mushrooms in the Central American Indian cultures. Image: © Paul Kroeger.

In the 16th century Spanish explorers in southern Mexico found that Indian tribes were using hallucinogenic mushrooms in rituals. They called these Teonanacatl, or 'flesh of God'. Using Teonanacatl gave the shaman a kind of second sight – they could see where stolen or hidden goods could be located. Anyone who partook of the fungus, as well as having visions, would dance and sing for a night and a day. There is even a record of fungi being used as an intoxicant at special festivals such as the coronation of a new ruler. In 1502 Diego Duran reported on the coronation of Montezuma II:

> All the company proceeded to eat the toadstools, a food which deprived them of their reason and left them in a worse state than if they had drunk much wine. Some became so intoxicated that they lost their senses and committed suicide. Others had visions during which the future was revealed to them.

This use of fungi in such ceremonies has been shown to date from around 500 BC, and a number of stone statues from this period have been recovered in Guatemala. Because of their shape it was initially believed they were part of a phallic

cult, but closer examination revealed they had small mushrooms carved around their bases. It is now believed they represent some of the earliest evidence of the importance of mushrooms to the indigenous peoples in this area.

Many species were still in use by the natives when Wasson was conducting his studies in the late 1950s. Wasson tried the fungi himself and reported that it was as if the walls of the room had disappeared, setting his soul free; he also saw angular, geometric patterns in many and varied colours which eventually coalesced into massive buildings. Wasson reported regular use of such fungi by the Amerindians.

Rene Robert has studied the work of painters under the influence of hallucinogenic fungi and reported that objects were generally elongated, with indistinct edges which were often surrounded by a halo. In addition, of course, the colours were wildly exaggerated.

The fly agaric (*Amanita muscaria*) is a well-known hallucinogenic species. There are reports of its recreational use in 18th-century Siberia; the fungus apparently produced great hilarity and giddiness, and eventually unconsciousness. Mention is also made of the fact that the hallucinogen passes unchanged into the urine. In other words, the urine of an imbiber is itself hallucinogenic, and this can be carried through

several humans before losing its potency. I have always thought how frustrating it must be to be the last person in line and upon drinking the intoxicant to find that the last of its potency had been used up with the previous individual!

It is claimed by some that the Viking Berserkers (a group of warriors who would lose all inhibitions in their fighting and not care what happened to them as they ploughed into the enemy) used fly agaric to work themselves up into the required state. However, most reports describe this fungus as exerting a peaceful and calming effect. It is worth noting it has the same effect on animals such as reindeer; dried samples used to be laid out as bait for reindeer and once the animals had consumed them they were much easier to catch.

Wasson devoted a great deal of time to work on fly agaric and its importance. He studied the ancient Hindu text, the Rig Veda, believed to date from 4000 BC, and reported his findings in *Soma, Divine Mushroom of Immortality*, published in 1968. Soma is described as a plant and also listed as a god, and appears to have many effects in common with fly agaric. Through comparing the texts and the characteristics of fly agaric Wasson became convinced that the two were one and the same.

According to John Allegro, a scholar of the Dead Sea Scrolls, fungi have influences on other religions as well. In *The Sacred Mushroom and the Cross* (1970) he argues that there was a fungus cult amongst the Israelites and that this related specifically to the fly agaric. His theories can be summed up in two quotes from the book:

[I]n the phallic mushroom, the 'man-child' born of the 'virgin' womb, we have the reality behind the Christ figure of the New Testament story.

…

If some aspects of the Christian ethic still seem worth while today, does it add to their authority that they were promulgated two thousand years ago by worshippers of the Amanita muscaria?

Naturally these ideas proved unpopular in some circles and fairly rapidly *The Mushroom and the Bride*, the Christian rebuttal, followed. One thing that neither book points out is the lack of records of fly agaric from the relevant location.

One final word on hallucinogenic fungi and their uses. The roots of the movement can be traced back to 1957, the year Wasson wrote an article for *Life* magazine called *Seeking the Magic Mushroom*. Thus Wasson can be seen as directly responsible for the widespread and generally (now) illegal use of the magic mushroom. Since the 2005 Drugs Act the magic mushroom (*Psilocybe semilanceata*) has been regarded as a Class A drug in the UK, placing it on a par with heroin.

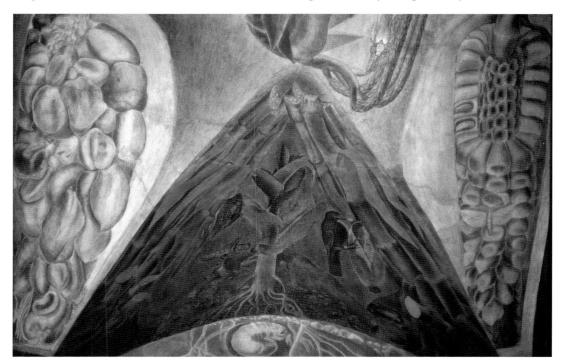

Left: Part of a fresco in the Abelardo Rodriguez Market in Mexico depicting corn infected by corn smut (*Ustilago maydis*). The infected corn is a valued foodstuff in Mexican culture. Image: © Dorothy McMeekin.

Right: Fungi feature in illustrations in many children's books. Perhaps the best known is John Tenniel's toadstool upon which sits a caterpillar smoking a hookah in Lewis Carroll's *Alice's Adventures in Wonderland*.

Fungi in literature, film and television

In literature, we see many instances in science fiction of fungi portrayed as monsters (see Chapter 7), but where would the murder mystery be without a liberal sprinkling of the death cap (*Amanita phalloides*)? There are in fact historical precedents for such flights of fancy. For example, the Roman emperor Claudius was thought to have been poisoned with (amongst other things) the juice of the death cap. The poisoning was carried out by his wife Agrappina and her son Nero. At that time another mushroom, Caesar's mushroom (*Amanita caesarea*), was so highly prized, rare and expensive that it was regarded as a delicacy fit only for the most high and mighty. It is reported that at a banquet after Claudius' death, Nero (now emperor) heard someone praising Caesar's mushroom as the food of the gods, and

responded: 'Oh yes indeed, it was fungi that made my father a god!' It was believed that when a Roman emperor died they became a god. The Holy Roman emperor Charles VI and Pope Clement VII are also thought to have been assassinated using the death cap.

Fungi in literature have other uses apart from poisoning. A famous example is that in *Alice's Adventures in Wonderland* by Lewis Carroll, written in 1865. Here we have the marvellous illustration by John Tenniel of a hookah-smoking caterpillar sitting atop a mushroom which is capable of transforming Alice in size. All told the Alice books are very 'trippy', and several authors have suggested that Lewis Carroll (actually the pen name of maths tutor Charles Lutwidge Dodgson) knew what he was talking about and that his writings were influenced by the partaking of hallucinogenic fungi himself.

In film and television we have many references to fungi – from the Fungoids of early *Dr Who* through to the horrific fungal diseases of *The X-Files*. One of the earliest film outings for Sax Rohmers' Fu Manchu was entitled *The Fungi Cellars* (1923). This features a cave lined with killer fungi; at one point we see two victims writhe in agony as they are attacked by the killer spores. In 2008 maverick film-maker Ron Mann premiered *Know Your Mushrooms* at the Vancouver International Film Festival.

The film was described by the *National Post* as 'Mann puts the fun in fungus. But more importantly, he puts the "us" into mushrooms.' The film 'invites you to enter the near secret world of edible fungi'. And, of course, how can we fail to mention the 2007 *Shrooms*? As the synopsis from Amazon.co.uk tells us:

> *A group of American students get a lot more than they bargained for when they ingest some magic mushrooms during a trip to Ireland. Although they're under the impression that they've acquired the best hallucinogenic money can buy, it's not long before they're convinced there's something sinister lurking in the woods nearby. Is there something picking them off, one by one? Or have the mushrooms begun to have an effect on them? Either way, it's the trip from hell, and it's a long time till morning...*

But for a truly bizarre mushroom cinematic experience try the 1963 Japanese epic *Matango, Attack of the Mushroom People* (also known as *Fungus of Terror*). Let's put it this way – it has recently developed something of a late-night cult following.

Fungi in art

In art, until about 1500 mushrooms seem to have been a relatively unimportant subject matter (notwithstanding the Palaeolithic paintings mentioned previously, whose purposes were probably not purely decorative). However, one surviving exception is the Apocalypse Tapestry found in Angers, France. This tapestry, dated to 1375, includes a small group of three mushrooms, immediately under the hind feet of the horse of one of the biblical Four Horsemen of the Apocalypse. Prior to this there had been a number of frescoes from Roman sites which included fungi of various species. In addition, there had been one or two other illustrations including Adam and Eve in the Garden of Eden standing next to the Tree of Knowledge of Good and Evil, or something that looks remarkably like fly agaric, *Amanita muscaria*.

From 1500 onwards fungi start to appear in pictures with great regularity. Initially many of these were brackets, which tended to be on trees as decoration. One of the most imaginative uses, however, must be the 1573 painting by Giuseppe Arcimboldo entitled *Autumn*. Arcimboldo was known for painting portraits in which fruit and vegetables were fitted together to form the image. In *Autumn* he uses a mushroom to represent an ear. The Baroque period of the latter half of the 18th century produced many works of art depicting tables of food of varying types, and fungi naturally appeared in many of these. From the 19th century onwards, gathering mushrooms became a popular subject for paintings,

Left: Prized edible wild mushrooms, such as the chanterelle (*Cantharellus cibarius*), are an economically important wild harvest. Image: © Chris Jeffree.

although it must be stressed that for several of these works it is only from the title we know fungi are involved as the mushrooms in question are not actually visible on them!

Nowadays the use of fungi in art is not uncommon; indeed, a number of paintings of fungi have been specially commissioned for use on stamps – surely an art form accessible to all.

Naturally there have also been many beautiful pictures of fungi produced for identification purposes and for illustrating field guides. If the reader is familiar with the work of Beatrix Potter only through Peter Rabbit then I strongly urge you to seek out some of her paintings of fungi – they are absolutely stunning.

Fungi as food and drink

Then there is the edibility of fungi (see Chapters 8 and 9). Most fungi are neutral when it comes to edibility – they will not harm you but you would probably get more pleasure from chewing cardboard. A small number are poisonous and will make you ill, and a tiny number are deadly. Similarly, a small number are edible and a few taste fantastic. Obviously, people's tastes vary so what some will rave over others will meet with a shrug of their shoulders, wondering what all the fuss is about.

Taken globally the figures and finances for mushroom cultivation are massive.

The most widely cultivated mushroom is *Agaricus bisporus*, the white button mushroom. It is grown in more than 70 countries, and the annual production has a value of around £1.5 billion. This one species, however, only accounts for 38% of the sales of cultivated mushrooms. We also need to take account of the wild-collected species (see Chapter 9), including mycorrhizal species such as *Cantharellus cibarius*, the chanterelle, and *Tuber aestivum*, the truffle. The most highly prized edibles are the white truffle (*Tuber magnatum*) and Caesar's mushroom (*Amanita caesarea*). In December 2009 the white truffle was being sold for £9,000 per kilogram. In December 2007

Below: A drawing of the Aztec Lord of the Underworld, Mictlantecuhtli, appearing behind a man eating hallucinogenic mushrooms, which Wasson published in his book *The Wondrous Mushroom.* Image: Courtesy of The Tina & R. Gordon Wasson Ethnomycological Collection Archives.

the world's largest recorded white truffle was sold at auction; it weighed 1.5 kg and sold for £165,000. Caesar's mushroom comes in as a poor relation at a mere £100 per kilogram.

Another important area of fungal food is mycoprotein (see Chapter 5), the leading UK form of which is marketed under the brand name Quorn®. Quorn® is actually *Fusarium venenatum*, a fungus found in soil. Originally grown as a protein substitute in 1967, it was finally given permission to be used as a human foodstuff in 1980, but it was not until 1994 that it was widely distributed. With the BSE issues of 1990 public confidence in the beef industry faltered and Quorn® sales saw a massive upturn.

We have already mentioned hallucinogenic fungi, but it should be remembered that alcohol is a yeast-based product, and the yeast (*Saccharomyces cerevisiae*) is of course a fungus. The earliest written records of brewing are from Sumerian clay tablets dating back some 6,000 years, although it is likely the process predates this by some considerable time. Yeast and other materials required for brewing were recovered from the Egyptian pharaoh Tutankhamun's tomb. Bread, which uses yeast to make it rise, has been known of for at least 8,000 years, although there have been significant changes in its nature over the millennia – for example, remains of early bread-eaters show much wear on their teeth due to the heavy, particulate consistency of the bread at that time.

Life saving fungi

Fungi are also responsible for saving the lives of countless millions through their use in antibiotic production (see Chapter 5). Medicinal uses of mould can be traced back to the Ancient Greeks and Romans, who used mouldy bread poultices to treat wounds; and medical textbooks from as early as 1640 cite moulds as being useful for various treatments. However, the modern discovery of antibiotics was made by Alexander Fleming in 1928. The story goes that Fleming was in a hurry to go on holiday and did not leave his lab as entirely clean as it ought to have been. When he returned he found that agar plates on which he had been growing bacteria had been contaminated by fungal spores. Fleming noticed there were clear

zones, where there were no bacteria, around the fungal colonies – the bacteria had been killed by a substance released into the substrate by the fungus (in this case *Penicillium chrysogenum*). Fleming isolated the active ingredient and named it penicillin – the first of many antibiotics which have since been discovered and isolated. Modern antibiotics can treat a range of diseases, though overuse has resulted in problems associated with resistance in some strains of bacteria.

Miscellaneous connections

With reference to the doctrine of signatures (where it is believed that like cures like), the stinkhorn (*Phallus impudicus*) is alleged to be an aphrodisiac and a cure for impotence. As its name suggests this fungus has a striking resemblance to human male genitalia!

Some species of *Ophiocordyceps*, parasites of insects (see Chapter 4), are believed to be performance enhancers and are reported to be used by some athletes. One common name for these is vegetable fly, and in China bundles of infected caterpillars were often sold as "summer plant – winter worm". At the time it was believed that the animal literally turned into a plant, well actually a fungus.

And so to one final thing: many argue that our modern interpretation of Father Christmas comes from the 1931 artwork of Haddon Sundblom, commissioned by The Coca-Cola Company. This definitely popularised and fixed the jolly fat man in red and white in our minds – but the elements had all been there before. Some argue that the colours in fact came from the red background and white spots of the fly agaric. The reindeer come in here too – reindeer were historically caught using fly agaric (as mentioned above), and one of the effects of taking fly agaric is a feeling of flying. So we have red and white, reindeer and flying. Add to this a mysterious figure living in the woods only appearing at certain times of the year and we have something that is relatively easily transformed into the modern version of Santa!

Amanita muscaria (fly agaric) species profile

There is the danger of a mycological elephant being present in the room. There is a species which we all know but no one will mention. Let me be the one to break this taboo. *Amanita muscaria*, the fly agaric. It's the mushroom with the red cap with the white spots on. You know the one, it's in virtually every fairy story you've ever seen illustrated.

It's an interesting mushroom. Fly agaric: agaric just means mushroom, and fly – well, it used to be dried and soaked in plates of milk where it would then be used as a fly catcher. The flies would come along and imbibe of the tainted milk and that would be the end of them. As well as intoxicating and killing flies it has been used as an intoxicant for humans too – historically it is a species much employed by shamans in religious and divinatory ceremonies.

In actuality it's an important mycorrhizal species of both coniferous and deciduous trees. Its natural distribution is almost universal over the northern hemisphere and it has also been introduced to the southern hemisphere along with trees in plantations. There is debate about whether the fly agaric is genuinely the same species the world over. Genetic analysis seems to indicate that there are distinct subspecies present – usually denoted by slightly different colours to the cap.

Fly agarics are the mushrooms commonly sold to accompany garden gnomes, and some even trace the physical appearance of Santa Claus (red and white) to the fly agaric. Of all species of fungi it is probably the one most commonly illustrated in non-specialist literature, and it must be the one most often sold as an ornament or a decorative item.

It's ubiquitous, it's widely recognised and it has a cultural pedigree second to none. Ladies and gentlemen, I give you, in all its gaudy glory, the fly agaric, *Amanita muscaria*.

Gordon Rutter

Image: © Patrick Hickey.

Main photos, clockwise from this page:

Dead man's fingers (*Xylaria polymorpha*); *Hypoxylon fragiforme*.
Images: © Chris Jeffree.

Witches' butter (*Exidia glandulosa*).
Image: © Peter Clarke.

Facing page inset photos:

Satan's bolete (*Boletus satanas*) (left).
Image: © Stuart Skeates.

Devil's fingers (*Clathrus archeri*) (right).
Image: © Martyn Ainsworth.

Chapter 7
Fungal Monsters in Science Fiction
Heather Kiernan

Fungal Monsters in Science Fiction

Heather Kiernan

In the minds of many people, fungi are associated with evil and witchcraft. This is reflected in the names of some, for example this is Devil's fingers (*Clathrus archeri*). Image: © Martyn Ainsworth.

The word 'monster' derives from a Latin root meaning 'to warn'. Monsters, real or imagined, literal or metaphorical, have exerted a dread fascination on the human mind for centuries. They attract and repel us, intrigue and terrify us, and in the process reveal something important about the darker recesses of our collective psyche. Monsters embody our deepest anxieties and vulnerabilities, and symbolise the mysterious and incoherent territory just beyond the safe enclosures of rational thought.

Science fiction has been at the crossroads of numerous disciplines. It has provided an incubator for imaginative minds to create visions that help us glimpse not only the future, but also something about ourselves in the present. Monsters have often been a part of those visions. Lord Robert May, population biologist and one-time president of the Royal Society, says of science fiction: 'At its best it is very provocative and forethoughtful, like Asimov's books – they think of questions that have since arisen such as intelligent machines.' Science fiction is more than just pulp fiction; at its core is the desire to understand humanity's place in the universe.

An easy target

Within science fiction fungi have repeatedly featured as monsters. They have come to embody evil and malevolence. Why have fungi often been viewed in this way, and what prevailing attitudes and fears in society have created fungal monsters? My focus here is science fiction literature but other forms and

Left: The sudden appearance of fungi, particularly when they are large, may be partly responsible for traditional ideas of an association with the underworld and evil forces. Image: © Kent Loeffler, Department of Plant Pathology and Plant-Microbe Biology, Cornell University.

media, such as poetry, film and video games, are also touched on.

There is nothing inherently bad about fungi. In fact, as all the chapters in this book make clear, the situation is quite the reverse – fungi are hugely beneficial to humans and the ecosystems of our world. However, it is easy to see how fungi could become the villain. An organism that can cause human fatality is straight away viewed with suspicion and fear. The fact that it is only a tiny minority of fungi that have the potential to kill humans is irrelevant, and all fungi seem to be tarred with the same brush. Names like the death cap (*Amanita phalloides*) and destroying angel (*Amanita virosa*) speak for themselves. The descriptions of poisoning conjure nightmarish visions of slow and painful death.

Poisonous fungi are bad enough, but there are other fungi that cause human disease (see Chapter 4). Another dramatic intrusion on human affairs is made by the fungi that rot our homes and possessions. Dry rot (*Serpula lacrymans*) rampaging through the timbers of a building, sending forth mycelial cords and fruit bodies (see Chapter 2), could be science fiction.

Fungi also lurk unseen – something Sir Arthur Conan Doyle took advantage of in his poem *Mind and Matter*, which begins:
Great was his soul and high his aim,
He viewed the world, and he could trace
A lofty plan to leave his name
Immortal 'mid the human race.
But as he planned, and as he worked,
The fungus spore within him lurked.

Below: The threatening nature of some fungi is seen in buildings ravaged by the dry rot fungus (*Serpula lacrymans*) where huge ominous strands of mycelium sometimes hang pendulously from collapsing timbers. Image: © Nia White.

Above: Satan's bolete (*Boletus satanas*) is poisonous, sometimes deadly. While it is harmful to humans – presumably its name reflects this – it is a mycorrhizal partner of tree roots feeding the plant water and nutrients (see Chapter 3). Image: © Stuart Skeates.

The extraordinary shapes and mysterious habits of fungi have provided an enduring source of inspiration for writers and poets in search of a metaphor for death and decay. Certainly fungi appear in folklore, as fairy rings (see Chapter 6) and the ingredients of potions. Shelley alludes to death and otherworldliness in his poem *The Sensitive Plant*:

> *And agarics, and fungi, with mildew and mould*
> *Started like mist from the wet ground cold;*
> *Pale, fleshy, as if the decaying dead*
> *With a spirit of growth had been animated!*

Fables about creatures that suddenly come to life are very ancient. For centuries, the sudden and rapid eruption of circles of mushrooms from the soil led people to believe that dark or terrible forces were at work.

Mycophobia

Right: Cultural attitudes to fungi vary hugely from mycophobia to mycophilia. This Croatian poster advertises a mushroom display organised by a local mushroom society in a country where fungi are appreciated as a valuable natural resource. Image: © Vladimir Jamnický.

Ethnomycology aims to uncover the sociology and historical practices associated with fungi. The discipline had its roots in the 1920s, when the amateur ethnobotanist R. Gordon Wasson and his Russian physician wife Valentina Pavlovna began to notice that a sharp dichotomy existed among cultures. In some, such as the Slavic culture, hunting for mushrooms was a tradition that had been in place since long before the Soviet era, and had provided a vital food source during periods of shortage. Mushroom picking certainly appears in several of the novels of

Tolstoy, and provided inspiration for the poetry of Pushkin.

In the Anglo-Saxon world, however, fungi had been largely shunned, associated with dung-heaps and poison; in Romantic poetry the smell of death still clung to them: 'from their very roots/Springs up a fungus brood, sickly, and pale/Chill mushrooms, coloured like a corpse's cheek', as Browning put it. The names were often used in a pejorative manner, as when the 18th-century novelist Laurence Sterne delighted in referring to the Scottish author and poet Tobias Smollett as 'the learned Smelfungus'. So Wasson made his first task coining the terms mycophilia and mycophobia to distinguish this difference in people's emotional attitude towards mushrooms.

Dangerous knowledge

The 19th century saw an explosion of scientific thinking and scientific exploration. The vision of the universe which traditional religion had championed was overthrown by the discoveries of science. From the age and history of the universe to the development of life, the old ideology yielded to a powerful assault of empirical evidence. Gothic novels seized on the scientific developments that had transformed the age. Part of their success stems from the

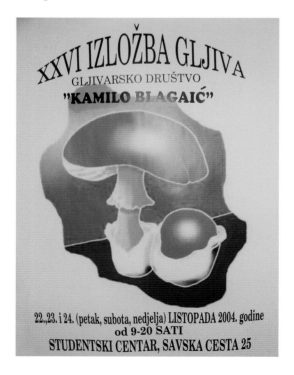

XXVI IZLOŽBA GLJIVA
GLJIVARSKO DRUŠTVO
"KAMILO BLAGAIĆ"

22.,23. i 24. (petak, subota, nedjelja) LISTOPADA 2004. godine
od 9-20 SATI
STUDENTSKI CENTAR, SAVSKA CESTA 25

way they encompassed the fears generated by technology, the chief agent of change.

A master of the Gothic horror tale, Edgar Allan Poe used death and decay as a central motif, mirroring the crumbling of society and religion under science. Fungi rising from decay permeate Poe's stories. In his short story *The Pit and the Pendulum* (1842), the prisoner condemned to death awakens in a dark chamber reeking of putrefaction and death, and Poe's protagonist notes 'the peculiar smell of decayed fungus' as a signifier of death.

A tale of sickness, madness, incest, and the danger of unrestrained creativity, *The Fall of the House of Usher* (1839) is perhaps Poe's greatest literary achievement and among his most popular horror stories. The narrator's first glimpse is of the 'melancholy' building, a 'mansion of gloom' whose 'principal feature seemed to be that of antiquity. The discoloration of ages had been great. Minute fungi overspread the whole exterior, hanging in a fine, tangled web-work from the eaves.'

Early science fiction also drew material, and metaphors, from a culture that could not have existed before the Industrial Revolution. Important developments in the fields of biology, physics, chemistry and metallurgy had a profound influence on this literary genre. In the prologue to a collection of Jules Verne's stories published under the title *Extraordinary Voyages* his publisher, Hertzl, wrote that the goal was 'to outline the geographical, geological, physical, and astronomical knowledge amassed by modern science'. No doubt the broad appeal of his work lay in the fact that the reader could learn something of science and the exotic locations and cultures of the world through the adventures of Verne's protagonists.

After descending miles into the bowels of the Earth, the three travellers in *Journey to the Centre of the Earth* (1871) discover a vast cavern filled with a deep subterranean ocean, surrounded by a rocky coastline covered in petrified trees and a 'forest of mushrooms'. Professor Lidenbrock remarks that he 'knew that the *Lycoperdon giganteum* [*Calvatia gigantea*] attains, according to Bulliard, a circumference of eight or nine feet; but here were pale mushrooms, thirty to forty feet high'. The reference to Bulliard would have made perfect sense to many readers at the time as he was the author of *Histoires des Champignons de la France* and made a significant contribution to the study of mycology.

Left: Dead man's fingers (*Xylaria polymorpha*) really can sometimes look like blackened fingers emerging from rotting wood.
Image: © Chris Jeffree.

Below: An etching by Édouard Riou for the novel *Journey to the Centre of the Earth* by Jules Verne. The picture illustrates the travellers discovering a giant cavern filled with gigantic prehistoric mushrooms (note the small figures for scale).

The writings of natural scientists, as well as more popular presentations of their ideas, found a substantial readership outside the scientific community. Writers, along with a notable slice of the educated public, eagerly ordered the latest scientific works the moment they were announced by the publisher. Charles Darwin developed perhaps the most prominent, controversial and far-reaching theory in all of biological science in *On the Origin of Species* (1859). Although Darwin's theory is disarmingly simple it is also easy to misconstrue. A correct, but rather one-dimensional, view of Darwin's theory was that individuals are locked in a pitiless struggle for existence, with extinction looming for those not fit for the task. It is perhaps not surprising that people did not always see the 'grandeur' that Darwin had. Often the theory was simply misunderstood. It was widely held that natural selection tended to encourage an almost infinite variability in form, and that all living things were constituted of protoplasm. Mixing these ideas with a Classical concept of hierarchy in nature, the *scala naturae* or the great chain of being, it seemed difficult to discount the terrifying possibility that human beings might mutate into lower-order beasts, or devolve even further down into the foundational slime – hence the importance of fungi as a menacing metaphor.

As the 19th century came to a close, efforts were made to come to terms with the unprecedented cultural trauma inflicted by the Darwinian revolution. Evolution has been one of the most important driving forces and imaginative resources for British science fiction writers. The iconoclastic novelist Samuel Butler had raised the possibility that machines could evolve 'mechanical consciousness' by means of natural selection. But it was in the great romances of H. G. Wells, the first man of letters to have had a modern scientific education, that the real anxieties of the new age were fully articulated as horror scenarios with a highly plausible evolutionary rationale.

Wells' short story *The Purple Pileus* (1896) is constructed around the premise that fungi have the power to alter human destiny. A timid and humiliated man named Mr. Coombes storms out of his home planning to end it all. As he walks through the countryside, he sees the pileus, a fungus of 'a peculiarly poisonous-looking purple', and eats it. 'They were wonderful things these fungi, thought Mr. Coombes, and all of them the deadliest poisons, as his father had often told him.' Here fungi are potential killers that turn out to have other powers over us.

Before the publication of *On the Origin of Species*, vivisection had been more or less tolerated. Darwin's ideas broke down the barrier between human and animal in the popular mind. Wells' *The Island of Dr Moreau* (1896) fed into the Darwinian debates of the time, shocking and horrifying many of its readers and reviewers. Wells effectively employed disturbing elements both to explore the implications of evolutionary theory, and to ask the reader to consider the limits of natural science and the distinction between men and animals. Fungi provide a sinister and unearthly feel to descriptions of the island.

In *The First Men in the Moon* (1901) Wells' growing interest in the political and sociological problems of the new century warns against imperialism, and a clash between civilisations. Imperialist culture, from the point of view of 'spacefaring', has often been used in science fiction as a commentary on colonialism on the Earth. Landing on the moon's surface the two heroes, Bedford and Cavor, find themselves on a planet with a thin, yet breathable, atmosphere, no animal life but abundant vegetation:

Ever and again one of the bladder fungi, bulging and distending under the sun, loomed upon us. Ever and again some novel shape in vivid colour obtruded. The very cells that built up these plants were as large as my thumb, like beads of coloured glass.

Overtaken by heat and hunger the two castaways begin to sample the native fungus. Intoxicated by the 'abominable' fungus, Bedford embarks on an argument likening their coming to that of Columbus in America 'to show the infinite benefits [their] arrival would confer on the moon'.

The following year saw the release of *Le Voyage dans la Lune* (1902), the screen's first science fiction story created by the ingenious French director Georges Méliès. A 14-minute masterpiece, the silent film's plot is a light-hearted satire criticising the conservative

scientific community of the time, and was inspired by both Jules Verne's *From the Earth to the Moon* (1865) and H. G. Wells' *The First Men in the Moon*. In the lunar underground kingdom, the scientists arrive at a mysterious grotto filled with enormous mushrooms of every kind. One of them opens his umbrella to compare its size to that of the mushroom, but the umbrella suddenly takes root and transforms itself into a mushroom, growing to gigantic proportions. Strange life forms then begin to emerge from under the mushrooms.

Invasion

Infiltration by aliens with a supra-human or animal advantage aroused imaginations and anxieties about hypothetical invasions by foreign powers in the years before 1914. Science fiction also began to project figures of fear. It is in Hodgson's *The Voice in the Night* (1907) that we are first introduced to the fungus monster and its offshoot, the fungus-possessed human.

Stranded at sea by a mist, a schooner is approached by someone in a nearby vessel. A man begins to tell a disturbing tale of being abandoned, together with his fiancée, by the sailors of the *Albatross* when it is badly damaged during a storm. After building a raft the couple drift for several days until they find themselves in a great lagoon. Discovering an abandoned ship, they soon become aware of odd patches of queer fungus studding the floors and walls of the cabins. No matter how often they scrape them away with carbolic, the fungi return to their original size within 24 hours. Deciding to make for shore they gradually become aware that 'the vile fungus, which had driven us from the ship, was growing riot'. Eventually the man and his fiancée discover that the fungus is invading their bodies. The final horror they face is their inability 'to withstand the hunger lust for the terrible lichen'. They discover that other humans on the island have apparently been entirely absorbed and become one with the strange fungal growth.

Hodgson applied the term 'abhuman' to the figures – formless, vile shapes of revulsion. As the fungus invades the body a person becomes 'the thing', a hybrid, neither solid nor liquid. 'Humans' are not human, and death is neither neat nor humanised, but a cauldron of decay and physical corruption.

John Wyndham's novels often deal with mutants and aliens arriving to supplant mankind, their telepathic powers an indicator of their evolutionary superiority. The message: we must either fight back or die out. Wyndham's *The Puff-Ball Menace* (1933) began to explore concepts and themes that would eventually lead to his more famous novels of the 1950s. A hostile country plants a dangerous fungus in Britain that breeds voraciously, and then kills the inhabitants.

Wyndham's post-apocalyptic world of *The Day of the Triffids* (1951) echoes Cold War fears that the release of infected insects, plants or animals could have serious effects on the food chain. *Fusarium*, a pathogenic fungus of many plants, was being used in biological weapons developed by the USA in the 1950s and 1960s. The idea of malevolent life still resonates today, with society increasingly concerned about the possible dangers of genetically modified organisms.

The concept that life came from outer space had been in the air since at least 1864, when Scottish mathematician and physicist William Thomson Kelvin told the Royal Society of Edinburgh:

> *The hypothesis that life originated on this earth through moss-grown fragments from the ruins of another world may seem wild and visionary; all I maintain is that it is not unscientific.*

In 1903, the Swedish chemist Svante Arrhenius removed the meteors from the equation. Instead, he proposed that individual spores could survive and travel through space, colonising any hospitable planet they landed on. The theory of panspermia was born.

In the science fiction film *Space Master X-7* (1958), Dr Charles Pommer, a scientist attached to the United States space exploration program, is bent on proving that life exists in other parts of the universe. To this end, Pommer sends a satellite, *Space Master X-7*, into orbit. On its return a canister containing samples of microscopic life from space is extracted and taken to Pommer for examination. When he opens it, Pommer finds a fungus he names

The brown puffball (*Bovista nigrescens*). John Wyndham, in one of his early novels – *The Puff-Ball Menace* – tells of a hostile country that plants a dangerous fungus in Britain. The fungus breeds rapidly and then kills the inhabitants. Image: © RBGE/Robert Unwin.

Right: The death cap (*Amanita phalloides*) is perhaps the most notorious deadly poisonous mushroom. Image: © Stuart Skeates.

Below: The nuclear age, and in particular nuclear bombs, ushered in fears about radiation. Mushrooms born of radioactive mutation feature in the Japanese film *Matango, Attack of the Mushroom People*. Image: © Science Museum/ SSPL.

'blood rust'. After dictating his findings into a tape recorder, Pommer begins to observe that the fungus is rapidly multiplying and might infect anyone coming into contact with it. Later, an incoherent Pommer telephones an urgent request for his lab to be destroyed. Rushing to Pommer's house, a colleague discovers the fungus covering Pommer's dead body and much of the lab. After listening to the tapes, he decides to set the house on fire before proceeding to the decontamination chamber. The documentary style of the film anticipates future terror stories of deadly organisms of extraterrestrial origin such as *The Andromeda Strain*.

New Dark Age

World War I had left Britain victorious but shattered. Man's increased reliance on science was both opening new worlds and hardening the manner by which he could understand them. Writers in the inter-war period, sickened by the horrors humanity had visited upon itself, began to see evolution as a good thing: the sooner a superior organism replaced mankind, the better.

Considered by many the successor of Poe, H. P. Lovecraft portrayed this potential for a growing gap in man's understanding of the universe as a potential for horror. In Lovecraft's sonnet cycle *Fungi from Yuggoth* (1930) an unidentified narrator acquires a mysterious book from an ancient bookseller. 'A book that told the hidden way/Across the void and through the space-hung screens/That hold the undimensioned worlds at bay.' The title is the name Lovecraft gives to a race of extraterrestrial fungoid creatures the narrator encounters that come from the planet Yuggoth to which he has unsuspectingly travelled. Later called the Mi-Go, these aliens and their planet reappear in Lovecraft's short story *The Whisperer in Darkness* (1931).

Mushroom clouds

The detonation of two nuclear bombs over Hiroshima and Nagasaki in August 1945 left a permanent scar both on Japanese history and on its psyche. The mushroom cloud ushering in the atomic age served as a symbol of the weapons themselves. It made an indelible mark on Japanese creativity, and the disturbing undercurrent would persist in popular culture for decades to come. The nuclear fallout over a vast area of the Pacific after the testing of a hydrogen bomb in March 1954 uncovered the hazards of radioactive substances, including genetic damage.

The appearance of a gentle, amphibious reptile of the Jurassic period, that had lain dormant in the depths of the ocean until a hydrogen bomb test transforms it into the gigantic ferocious monster Godzilla, was the first of many science fiction films born of radioactive mutation. Godzilla reflected the nuclear anxiety of both the Americans and the Japanese.

Left: Excavated mummified caterpillar with a fruit body of *Ophiocordyceps sinensis*. Image: Reproduced with kind permission of Elsevier. © Daniel Winkler.

The grotesque mutations caused by the mysterious 'matango' mushrooms inevitably recall the unspeakable nuclear trauma of Hiroshima and Nagasaki. Loosely based on *The Voice in the Night*, the film *Matango, Attack of the Mushroom People* (1963) has a Japanese yacht blown off course by an unexpected storm. The crew and passengers find themselves on the shoreline of a deserted island, but the shortage of food looks set to be short-lived, as giant clumps of mushrooms grow in abundance on the lush tropical island. The marooned visitors soon discover the hull of a stranded research vessel whose crew appeared to have been studying the effects of radioactive fallout.

The discovery of the ship's log sheds some light on the mystery, warning of the giant fungus's damaging effect on the human nervous tissue if eaten, as does the appearance of a green slimy mutated mushroom man at one of the portholes. But by this stage it is too late, as the passengers descend into paranoia and psychosis under the influence of the hallucinogenic fungus. As the mushrooms begin taking over the island, the mutated fungoid remains of the crew of the previous wreck close in. The only survivor escapes to

Left: Yellow brain fungus (*Tremella mesenterica*). The mycelium of this fungus is parasitic on other fungi. Image: © Patrick Hickey.

Above: Witches' butter
(*Exidia glandulosa*).
Image: © Peter Clarke.

Obsession and corruption

Come Into My Cellar (1962) is a short story by one of the most revered writers of speculative fiction, Ray Bradbury. Later made into an episode of the TV anthology series *The Ray Bradbury Theatre*, the story tells of the many wondrous things that small boys can order from mail order catalogues. Hugh, a worried father, begins to believe that his son's mushroom garden is taking over the neighbourhood. Hugh's best friend skips town without explanation, his neighbour claims she's Earth's first line of defence against an alien invasion, and his son spends an inordinate amount of time in the basement. Hugh's speculations lead him to take a closer look at the mail-order mushrooms many of the town's boys, including his son, are growing in their cellars.

In *The Trouble with Lichen* (1960), Wyndham explores what happens when two scientists simultaneously identify a substance that slows down the aging process. Both of them realise that such a discovery will give rise to exploitation and widen the gap between rich and poor, and each ponders how best to use it for the betterment of mankind. One decides to suppress it, though he is not above using it on himself and his family. The other decides that she trusts women more than men, and in order to tap into the influence that women have over their husbands, opens an exclusive and expensive beauty salon, which is frequented by the wives of MPs and business magnates. In a global anti-aging market that is estimated to be worth approximately US$60 billion, the issues surrounding the commercial exploitation of science-based cosmetics are even more urgent now than they were when the book was written.

civilisation and tells his story from a padded cell, as mushrooms start to sprout from his face.

The imagery of atomic explosion and post-apocalyptic salvation also found expression in much of Japan's subculture. The world of *Nausicaä of the Valley of the Wind*, a series of 'manga' (graphic novels) published between 1982 and 1994, is an ever-expanding radioactive wilderness that threatens the pockets of humanity who survived 'The Seven Days of Fire' huddled in small enclaves across the continents. The Fukai, a toxic forest of fungal plants, whose spores are poisonous to humans, covers much of the Earth's surface.

Feeding on the pollutants of the former human civilisation, the Fukai expands, enveloping the outposts of mankind and consuming them. Giant mutant insects are now the dominant form of life. One of the islands of humanity, known as the Valley of the Wind, is protected from the spores by strong winds from the sea. Delicate spores dangle from gossamer webs, waiting for the unsuspecting visitor to inadvertently touch them thereby releasing their deadly store of radioactive gas.

Right: *Nectria cinnabarina*.
Image: © Chris Jeffree.

Some fungi produce light by the process of bioluminescence, as seen in *Omphalotus nidiformis* from Australia. It is still unclear if there is any purpose to this, and it may simply be a by-product of metabolic processes. Nevertheless, it is a wonderful and perhaps eerie phenomenon. Image: © Ray and Elma Kearney.

Relentless spread

With the end of the Cold War, terrorism, pandemics, and weapons of mass destruction continue to fuel fear and insecurity in a seemingly apocalyptic world. Fungi carry on weaving their filamentous hyphae through the literary imagination. Examples include John Lanchester's darkly comic novel *The Debt to Pleasure* (1996), in which the protagonist's murder spree culminates in a carefully planned mushroom poisoning, Virginia Smith's romantic novel *Murder by Mushroom* (2007), Stephen King's short story *Gray Matter* (1973), John Brosnan's 1985 novel *The Fungus*, and Brian Lumley's *Fruiting Bodies* (1989), which won a British Fantasy Award.

Sentient fungal creatures appear in Jeff VanderMeer's acclaimed fantasy novel *Shriek: An Afterword* (2006). In the imaginary city of Ambergris, we meet the inscrutable, subterranean 'Grey Caps', Ambergris' original inhabitants, which yearn to transform the city with a fungal-based technology. Sentient fungi have also featured in numerous video games since the 1990s. However, the mutable fungus creature, independent of a human host body, can be traced back to the 'Myconids' of the role-playing game *Dungeons and Dragons* (1974). The Myconids provide a rare example of the positive portrayal of fungi as social and hardworking beings.

Whether we like it or not, and regardless of the reality, fungi have been consistently cast in the role of monster in science fiction. There are obvious reasons why this should be: poisoning, consumption and decay. The otherworldliness of fungi, stemming from their strange forms and rapid appearance and disappearance, predisposes them to

Above: *Microglossum viride*, an earth tongue.
Image: © Patrick Hickey.

being viewed with suspicion. Paradoxically, it is perhaps the apparent harmlessness of fungi that makes them such good monsters. Sylvia Plath's poem *Mushrooms* is a striking example of this perception of the horror behind apparent innocence. The poem offers a gradual disclosure of the threat of destruction posed by small, seemingly harmless creatures. Plath chooses mushrooms to illustrate the general failure to see beyond appearances. There is a quiet relentlessness about fungi which, in the right hands, is terrifying. In the final stanza of *Mushrooms* Plath leaves us in no doubt that the fungal monster will get us:

> *We shall by morning*
> *Inherit the earth.*
> *Our foot's in the door.*

Fungal monsters will continue to thrill, terrify, entertain and inspire us. We catch glimpses of these creatures all around: covering rocky coastlines, looming in the lunar underground, or as 'Myconids' able to communicate by emitting spores. Mythic creatures occupy a place in our culture that answers both our fears and our desires about what it is to be human, and therefore they will always have something to tell us.

Armillaria mellea (honey fungus) species profile

The honey fungus is a serial killer of trees; it invades our gardens with vast flushes of toadstools. But did you know that it glows in the dark?

Armillaria mellea is bioluminescent – a living organism that produces light. In this species it is only the mycelium that glows, but in some other fungi the fruit body also glows. Noted in 312 BC by Aristotle, fungal bioluminescence has been observed and utilised by many cultures, and is often referred to as 'foxfire' or 'will o' the wisp'. Soldiers in World War I attached bioluminescent wood to their helmets to avoid bumping into each other at night, and pieces of glowing wood and branches are worn by the Bayaka pygmies during 'forest spirit festivals'.

My own investigations of this remarkable phenomenon in honey fungus have shown that the intensity of the light produced fluctuates through time. In time-lapse film the glow has an eerie pulse and movement to it. It is tempting to speculate about the purpose of this fluctuation. Perhaps it attracts insects which feed on the fungus and unwittingly disperse it? Rather more mundanely, it could just be a by-product of metabolism.

Armillaria mellea is a white-rot fungus which can wipe out entire forests (the numerous dead trees left behind become known as 'ghost forests'). The genus to which it belongs is widespread around the world and comprises around 40 species which vary in physical form and ability to cause disease.

The mycelium of *Armillaria* produces specialised root-like structures known as rhizomorphs or 'bootlaces', which form extensive networks. Rhizomorphs contain melanin (the same pigment found in our skin and hair) in their tough, leathery outer layer, which acts a bit like a rain coat preventing entry and loss of water and nutrients. Rhizomorphs are remarkable structures composed of thousands of individual hyphae, acting as 'super-highways' capable of transporting nutrients and water over several metres. The largest organism currently known on the planet is an individual rhizomorph network of *Armillaria ostoyae* in Oregon, USA. It is thought to be around two thousand years old and covers more than 3 square miles; the mycelium is estimated to weigh more than 600 metric tons!

Patrick Hickey

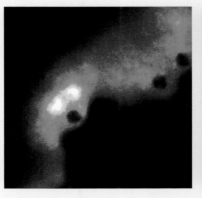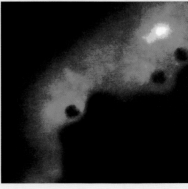

Fluctuations in bioluminescence within a colony of *Armillaria mellea*. Images six hours apart.
Images: © Patrick Hickey.

Main photos, clockwise from this page:

Lycoperdon perlatum;
Daedaleopsis confragosa;
chanterelle
(*Cantharellus cibarius*).
Images: © Chris Jeffree.

Facing page inset photos:
Shiitake (*Lentinula edodes*)
fruiting on a sawdust block
(left); oyster mushrooms
(*Pleurotus ostreatus*) (right).
Images: © Ann Miller,
www.annforfungi.co.uk

Chapter 8

Growing Edible Fungi

Patrick Hickey

Cautionary notes:

- be certain mushrooms are not poisonous
- consult an expert if unsure
- if in doubt don't eat
- the first time you try a new mushroom only eat a little

Growing Edible Fungi

Patrick Hickey

Right: Grey oyster mushrooms (*Pleurotus ostreatus*). Image: © Ann Miller, www.annforfungi.co.uk

Below: The familiar button mushroom (*Agaricus bisporus*) is probably the most widely cultivated mushroom species, although in some parts of the world other species are more commonly grown. Image: © RBGE/Max Coleman.

Gourmet mushrooms are highly nutritious delicacies enjoyed by cultures around the world, and many are also famed for their medicinal properties. Edible mushrooms are available harvested from the wild, or may be cultivated. This brief review focuses on those species that can easily be grown by the home mushroom cultivator and outlines some basic techniques of mushroom cultivation.

The 'normal' mushroom that we eat is *Agaricus bisporus*, and the earliest record we have of its cultivation is from 1707, courtesy of French botanist Joseph Pitton de Tournefort. Today cultivation takes place in more than 70 countries and the annual production is more than 1.5 billion kilograms, estimated to be worth £1.5 billion. However, this species only accounts for some 38% of the sale of cultivated mushrooms. It is one of the easiest to produce but is also among the blandest in terms of taste.

According to the International Society for Mushroom Science, *Pleurotus* species, the oyster mushrooms, account for 25% of mushroom (in the broadest sense) cultivation, with China leading the rest of the world by some considerable margin in terms of quantities produced and consumed. Also big in Asia are the paddy straw mushroom *Volvariella volvacea* (16%) and *Lentinula edodes* or shiitake (10%). There are many techniques for cultivating fungi, and a wide range of kits are now commercially available

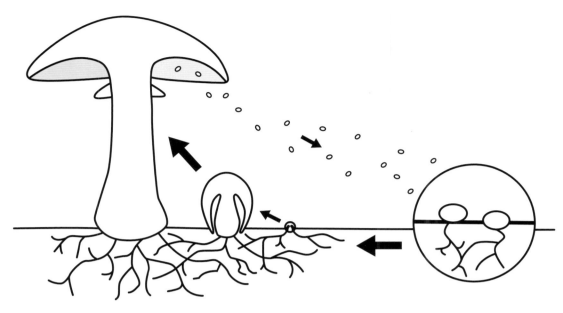

Left: Diagram illustrating the mushroom life cycle. Spores are released from mushroom gills and may be carried long distances by wind. When spores encounter a suitable substrate for growth, they germinate and grow into mycelium. Eventually, the mycelium forms pinhead sized aggregations termed primordia which develop rapidly over a few days into fruit bodies. Illustration: © Patrick Hickey.

to the mushroom enthusiast. Three different mushroom growing methods will be outlined here: compost patches, indoor cultivation in plastic bags/boxes, and outdoor cultivation on logs.

The main body of fungi is termed mycelium (see the Introduction), which, under the right conditions, gives rise to the reproductive fruit bodies we know as mushrooms. A basic understanding of the mushroom life cycle is helpful for successful cultivation. Most cultivated fungi are saprotrophs, that is, they grow on dead organic matter (see Chapter 2). Fungi are therefore the ultimate recyclers and many waste materials, for example woodchips, cardboard and straw can be transformed into edible mushrooms. Certain species of edible fungi form intimate associations with plant roots (see Chapter 3), and these are known as mycorrhizal fungi. The mycelium of a few mycorrhizal fungi can be cultured; however, without the plant host, the fruit bodies (mushrooms) are never produced. Attempts have been made to grow chanterelle and truffle fungi by inoculating host trees, but such methods require careful management and years of patience.

Far left: Shiitake mushrooms (*Lentinula edodes*) can be cultivated on hardwood logs. Image: © Patrick Hickey.

Left: Mycorrhizal fungi like chanterelles (*Cantharellus cibarius*) cannot be cultivated without its mycorrhizal partner. Image: © Patrick Hickey.

Right: Oyster mushrooms (*Pleurotus ostreatus*) fruiting from pasteurised wheat straw in a plastic box. It is important to ensure adequate drainage to prevent the growth substrate becoming water saturated. Image: © Patrick Hickey.

Home cultivation

The basic principle for growing all fungi is to obtain a pure culture of mycelium (spawn) and to inoculate a suitable growth substrate. Different strains are selected for characteristics such as mushroom appearance, flavour, substrate preference and optimal climate conditions. When deciding what type of fungi to grow, it is useful to consider space requirements and what substrates are locally available, for example logs, woodchips, straw or manure.

There are many places in which mushrooms can be grown. A garage, shed, basement, caves or simply a shady corner in the garden all offer possibilities. Mushroom patches and inoculated logs should be protected from direct sunlight, wind and extremes in temperature. The substrates, growing methods and conditions required for a selection of species are listed below.

Species/ Common Name	Growing Method and Substrates	Colonising Temperature	Fruiting Temperature	Comments
Agaricus bisporus Button Mushroom	Bed/patch Compost/manure Casing required	24–25°C	18–21°C	Great for beginners, numerous kits available.
Coprinus comatus Shaggy Ink Cap	Bed/patch Compost/manure, leaf mulch Casing required	22–27°C	16–18°C	Good compost heap coloniser. Fruit bodies best eaten when young, before they literally 'melt'.
Hericium erinaceus Bearded Tooth	Sawdust mixture in plastic bags Hardwood logs	22–28°C	18–20°C	Very interesting texture and a delicate flavour which resembles lobster.
Lentinula edodes Shiitake Mushroom	Sawdust mixture in plastic bags Hardwood logs	20–25°C	15–20°C	Fruiting may be induced by soaking logs in cold water for 1–2 days.
Lepista nuda Wood Blewit	Bed/patch Compost, leaf mulch and straw	18–24°C	15–20°C	Mushrooms must be thoroughly cooked before eating.
Pleurotus ostreatus Oyster Mushroom	Hardwood logs Pasteurised straw in plastic bags	22–28°C	12–17°C	Recommended for beginners. Rapid growing, usually crops within 4 weeks.
Stropharia rugoso-annulata Giant Stropharia	Bed/patch Hardwood chips, straw, cardboard	20–28°C	10–20°C	This fungus is great for colonising woodchip beds and borders and will not damage plants.

Spawn

Spawn is a pure culture of mycelium generated on a carrier material such as wooden plugs, grain or sawdust. The spawn is used to inoculate a substrate of choice (straw, logs, etc.), thereby distributing live mycelium uniformly throughout the substrate and allowing the fungus to become quickly established. Spawn is often produced in ventilated plastic bags made from heat-resistant plastic so they can be sterilised by pasteurisation or in an autoclave. Vented plastic bags and boxes can be readily sourced from internet-based suppliers. For someone starting out in mushroom growing the most effective way is to obtain fresh spawn from a reputable supplier. For serious growers it becomes more cost effective to produce your own spawn. Fresh spawn can be refrigerated but has a limited shelf-life of only a few weeks.

Indoor cultivation

A humid growing area is required for indoor cultivation. Large plastic storage boxes or plant propagators may be adapted into mushroom fruiting chambers. Electric heat mats can be used to maintain a constant temperature, but care should be taken that mycelium and substrate do not come into direct contact with the heat source (e.g. by using a wire mesh to support substrate containers above the heat mat). Humidity can be maintained by suspending growth containers above a layer of moistened perlite (a volcanic soil additive) or through use of water-retaining gel. Mist can be generated using an ultrasonic humidifier (sold in garden centres) or using sterile water in a spray misting gun. Unlike plants, fungi do not require light for growth, and the mycelium will grow happily in complete darkness. However, for abundant fruiting most species require low levels of natural light or fluorescent grow lights.

Left: Inoculation of woodchips using grain spawn. After adding the grain spawn the bag is sealed and gently shaken to distribute the spawn evenly. The bags are manufactured using heat resistant plastic allowing steam sterilisation. The white patch is a filter which allows gas exchange. Image: © Patrick Hickey.

Left: Cultivation of oyster mushrooms (*Pleurotus ostreatus*) within plastic bags in Vietnam. Image: © Dao Thi Van.

Pest control

Contamination of substrates with pests in the form of unwanted fungi and bacteria is an ever-present problem. Such contamination can be avoided by using efficient sterile techniques and by careful planning. Alcohol or disinfectant should be used to wipe down work areas. Competing fungi such as *Trichoderma* species and *Diehliomyces microsporus* (see Chapter 10) spread quickly and wreak havoc in mushroom growing rooms. Other common fungal pests on straw-based substrates include *Aspergillus* and *Penicillium*. Bacteria sometimes cause problems on grain spawn, particularly when there is poor air circulation and excessively humid conditions.

Slugs and snails will harvest your mushrooms if they get to them before you do. A single snail can devastate a cluster of developing mushrooms. Birds and foraging animals may eat grain spawn and can be discouraged by covering cultivation areas with fleece or polythene sheeting, and by raising logs above ground. Insect pests such as woodlice, black flies, gnats and mites that live in soil can all devour mycelium. It is good practice to identify 'clean' areas and regularly dispose of spent substrates. Chemical pesticides are generally not recommended and should never be used directly on mushroom beds/logs as they can kill the mycelium or even be taken up and concentrated within the mushroom. Sticky traps (sheets of plastic or paper covered with adhesive glue) can help to control insects and mites.

Safety first

It is unwise to eat any type of mushroom unless you are confident you know what it is (see Chapter 9); if you are unsure, refer to an expert. In rare circumstances, contamination of substrates by rogue species can result in non-edible fungi fruiting alongside edible mushrooms. As with many foodstuffs, some people will be intolerant to certain foods, therefore the first time you eat a particular type of edible mushroom it is wise to eat only a small amount. Mushrooms shed millions of spores each day. Like pollen, spores are a potential health risk, particularly for those prone to allergy and with underlying respiratory problems (see Chapter 4). Spore-less, or very low spore-producing, varieties of *Pleurotus* have been developed to combat this problem.

Cultivation methods
Compost patches

Mushroom patches based on composted substrates can be established in many different forms, for example trenches, pits, mounds, trays, shelves, troughs or raised beds. It is also possible to inoculate garden compost heaps in the spring, and plentiful crops of mushrooms should appear later in summer/autumn. It is important that spawn is introduced to compost after the initial phase of decomposition, however, as excessive heat produced during this phase can kill off mycelium. Compost does not have to be sterilised, but pasteurisation is usually employed for commercial-scale operations.

The procedure for preparing an outdoor mushroom patch involves digging a shallow trench, spreading out a layer of compost/substrate, followed by a layer of spawn, and covering with more compost/substrate.

Below: Straw mushrooms (*Volvariella volvacea*) growing on rice straw in Vietnam. Image: © Dao Thi Van.

Left: Modern cultivation techniques employed at Fungi Perfecti (USA). The bearded tooth (*Hericium erinaceus*) fruiting in specialised growth rooms utilising under-floor heating, humidifiers and ventilation. Image: © Patrick Hickey.

Usually mushroom patches are topped by a casing layer of soil. Casing is an essential part of mushroom cultivation for species such as *Agaricus* and *Coprinus*. A thin layer, around 10–20 mm depth, of peat moss (usually supplemented with chalk or limestone flour to balance the pH) provides a moist protective layer and helps induce mushroom formation.

Plastic bag culture

Oyster mushrooms (*Pleurotus*) are popular in supermarkets and are some of the fastest-growing cultivated mushrooms – fruit bodies may be produced in as little as two weeks after the initial inoculation with spawn. The pink oyster and yellow oyster mushrooms are closely related tropical varieties that require slightly higher temperatures than the grey oyster mushroom *Pleurotus ostreatus*. Plastic storage crates or laundry baskets with holes in the sides may be used as inexpensive containers for growing oyster mushrooms. The mushrooms can also be grown on logs using the same technique as for shiitake (see below). The following procedure outlines cultivation of *P. ostreatus* on pasteurised wheat straw within special ventilated plastic bags.

Clean, finely chopped wheat straw is pasteurised by steeping in boiling water for around one hour, followed by thorough draining. (Pressure cookers, microwaves or vegetable steamers are alternative methods of pasteurisation.) The straw is placed in clean, ventilated plastic bags, and allowed to cool. The moist, pasteurised straw is then inoculated

Below: Pink oyster mushrooms (*Pleurotus djamor*) growing on wheat straw in a plastic bag. Image: © Patrick Hickey.

using an approximately 1:10 ratio of spawn to substrate (e.g. 100 g of spawn is enough to inoculate 1 kg of substrate). After the spawn has been added, the bags are immediately sealed, shaken to mix in the spawn, and then incubated in the dark at 20–25°C for 2–6 weeks. A brief cold shock and spray misting can induce fruiting; holes are cut in the sides of the bags to allow mushrooms to emerge. Bags are placed in a fruiting chamber with a light source, and with humidity maintained by regular misting. Many different species can be grown using bag-culture techniques.

Growing shiitake on logs

Growing mushrooms on logs is recommended for more ambitious and patient cultivators as the process generally takes one to two years to produce mushrooms. The process involves selecting freshly cut hardwood logs, drilling holes in them and inoculating with spawn in the form of mycelium-impregnated wooden dowels. The best types of wood for shiitake (*Lentinula edodes*) cultivation are oak, birch or alder. The optimal diameter is around 10–20 cm, as the mycelium grows best in sapwood, and for ease of handling logs should be cut to 1 m lengths.

The trees used to provide logs should be felled towards the end of winter, cut and rested for up to eight weeks prior to inoculation in the spring. The exposed ends of logs may be sealed with wax to discourage competing ubiquitous fungal species. Ideally about 30 holes should be drilled per log, distributed evenly. The spawn plugs should then be carefully inserted into

Above: Spawn plugs of shiitake mushroom (*Lentinula edodes*). Image: © Patrick Hickey.

the holes, the diameter of which should be a close fit. The hole should be drilled slightly deeper than the length of the plug, leaving a small unfilled space centrally within the hole. Sites of inoculation are then sealed using molten food-grade cheese wax that has been heated in an old pan. Extreme care should be taken as boiling wax is highly flammable and will cause severe burns. A turkey basting pipette, or a spoon, ladle or paintbrush, may be used as an improvised wax application tool.

Following inoculation, logs are stacked crossways to optimise space and numbered in order to keep track of strains and the date inoculated. Inoculated logs should be protected from direct sunlight, strong winds and severe frosts and can be covered with tarpaulin or fleece to retain moisture. Fruiting can be induced by shocking logs by immersing them in a bath of cold water for two days. Left to nature, logs tend to fruit in autumn and spring, usually after rainfall, a change in temperature or other physical stress. Logs inoculated in the spring may begin fruiting the following autumn, and logs may continue to fruit for many years.

Mushroom growing can be an extremely rewarding and novel hobby. There are few hobbies that allow simultaneous recycling of waste whilst producing delicious treats for the table!

Lentinula edodes (shiitake) species profile

Shiitake mushrooms are widely consumed around the world, particularly in China, Japan, Thailand and Korea. Records show they have been cultivated for thousands of years, by inoculating hardwood logs with spores or mycelium. The name comes from 'shii', a type of tree popularly used to cultivate the mushrooms in China, and 'take' meaning mushroom.

I grow my own shiitakes on alder and birch logs in a shady spot in the garden. Left to nature, my shiitake logs fruit twice a year – in spring and autumn. One batch of logs which were inoculated with mycelium spawn over ten years ago are still active. The logs have almost disintegrated, yet regularly produce plentiful crops of mushrooms!

Shiitake are popular in Asian cuisine, usually sautéed, steamed or boiled in clear 'miso' soup. Shiitake should always be cooked before eating. I prepare shiitake by removing the tough stems and slicing the cap into thin sections. Shiitake are sold both fresh and dried. The sun-drying process converts ergosterol to vitamin D and also breaks down proteins into amino acids, giving the mushrooms a more meaty flavour. The Japanese call this the 'umami' – meaning taste/flavour. The glutamates produced are in fact natural equivalents to the flavour enhancer monosodium glutamate (MSG). Some people will have eaten shiitake without even knowing it. I have noticed that in recent years the dried button mushrooms included in some dried noodle products sold widely in British supermarkets have been replaced with shiitake!

Shiitake are claimed to have medicinal properties, and are used in traditional Chinese medicine. A compound isolated from shiitake mushrooms, called lentinan, has been examined for its anti-cancer properties, and other components are said to lower cholesterol and help reduce blood pressure.

Patrick Hickey

Shiitake mushrooms fruiting from a sawdust block.   Image: © Ann Miller, www.annforfungi.co.uk

Main photos, clockwise from this page:

Shaggy parasol
(*Chlorophyllum rhacodes*).
Image: © Peter Clarke.

Shaggy parasol
(*Chlorophyllum rhacodes*);
morel (*Morchella esculenta*).
Images: © John Wright.

Facing page inset photos:

Wood blewit (*Lepista nuda*)
(left and right).
Images: © John Wright.

Chapter 9
The Fungal Forager

John Wright

Key points for the fungal forager:

- know and follow the law
- pick only what you will study or eat
- avoid causing habitat damage
- be certain mushrooms are not poisonous before eating
- consult an expert if unsure
- if in doubt don't eat

The Fungal Forager

John Wright

Cep (*Boletus edulis*).
Image: © John Wright.

I am always a little envious of my friends who lead butterfly hunts, bird-watching expeditions and lichen-spotting forays – they never have to answer the question I have been asked several thousand times on my fungus forays: 'Can you eat it?' I have, however, come to terms with the narrow view this question implies for it provides people with a door into an area of the natural world that other disciplines lack. Those who go mushroom hunting bring away more from a foray than their tea; they learn something of how nature works and come to value its beauty and complexity. Indeed, many have gone on to make mycology a lifelong passion.

Of course the second most common question I am asked is a general form of the first: 'How can you tell if something is edible or poisonous?'

Identification

I expect that most people are hoping for a fairly straightforward answer; perhaps that fungi with dark gills are edible or that mushroom-shaped fungi growing on wood will do you no harm or that anything green is poisonous. It is at this point I always disappoint my audience. None of these statements are true and certainly none of the old wives' tales about peel-able caps or discolouration of silver spoons. There is *no* rule-of-thumb, not one. Of course there *is* an answer to the question: study the fungi for 10–20 years and all will become (reasonably) clear. This, however, is not a popular response. But are there, comes the plea, any short-cuts?

Of course you do not have to be an expert mycologist in order to pick a few mushrooms any more than you need a degree in botany to collect a pot of blackberries or sloes. All you need is to learn to identify a handful of common species – preferably distinctive and tasty ones.

Cep (*Boletus edulis*), chanterelle (*Cantharellus cibarius*), parasol mushrooms (*Macrolepiota procera*), shaggy ink caps (*Coprinus comatus*), horse (*Agaricus arvensis*) and field (*Agaricus campestris*) mushrooms, puffballs (*Lycoperdon* species) and hedgehog mushrooms (*Hydnum* species) are all easily found and easily identified, and the beginner could do worse than concentrating on this short list. But even with these you will need to learn the basics of identifying the larger fungi. It would also be wise to learn any 'lookalikes' that may cause confusion, and certainly to familiarise yourself with the commoner and the more serious poisonous species.

There you are, sat in your kitchen with a basket full of mysterious fungi, trying to match the ten species in your collection with the 1,500 plus pictures in your book; where on earth do you begin?

In fact you will have begun already. The first stage in identifying a specimen is to note its habitat. Some fungi can be found almost anywhere but most are quite fussy about where they will grow. Sometimes it is just a preference for woodland or grassland but often it is more specific. Many of the real gourmet fungi such as ceps, chanterelle, horn of plenty (*Craterellus cornucopioides*), truffles (*Tuber* species) and the saffron milk cap (*Lactarius deliciosus*) have an association with trees (mycorrhizal; see

Chapter 3). This means that an ash wood will produce a very different array of fungi from a beech wood. Every mushrooming guide will tell you the sort of habitat in which a fungus will be found. So it might say 'Short grass with moss' or 'With oak or beech'. These are important clues and it is essential to make a note of where a particular fungus was found – you could take a photograph of your find *in situ* or just pop a leaf from the nearest tree in with your specimen as an aide-memoire.

There are other observations best made on really fresh material which are worth making at the time of collection. One characteristic critical

Above: Hedgehog fungus (*Hydnum repandum*). Image: © John Wright.

Far left: Chanterelle (*Cantharellus cibarius*). Image: © John Wright.

Left: A basket of mixed (not necessarily edible) fungi collected on a foray at Dawyck Botanic Garden in the Scottish Borders. Image: © Chris Jeffree.

Above: Yellow stainer (*Agaricus xanthodermus*) is poisonous, with symptoms including sweating and stomach cramps. Image: © John Wright.

Above right: Yellow stainer living up to its name. Image: © Patrick Hickey.

for some species is a colour change when the surface or flesh is damaged. The poisonous yellow stainer (*Agaricus xanthodermus*) will bruise chromium yellow when the edge of the cap or base of the stem is rubbed. And the lemon yellow flesh of the edible (when cooked!) scarletina bolete (*Boletus luridiformis*) turns an instant and startling dark blue when cut.

Another character is smell. There are easily 20 species that can be identified by smell alone but most fungi have *some* smell and this may be another, and important, aid to identification.

While it is never wise to collect a large number of any one species on the off chance

Right: Scarletina bolete (*Boletus luridiformis*) is edible when cooked, but it has been known to cause gastric upsets. Image: © John Wright.

that it may be edible, it is best to have at least two fruit bodies to aid identification. Generally a mature one and a young one will provide all the characteristics you need.

Unless you know for sure what you are collecting it is very important to carefully lever your find out of the ground in order to preserve as much of the fungus as possible. The base of the stem will often provide essential clues to the identity of a fungus; for example, the death cap (*Amanita phalloides*) has a bag at the base and the yellow stainer bruises intense yellow inside the flesh at the bottom of the stem. If the fungus has been cut these characters are lost for ever. Identification characteristics are often very subtle so do handle your specimen with care. Your book may say something like 'fragile ring on stem'. If you have been using the stem as a handle then all you will get is fingerprints.

Specimens should be kept in open plastic boxes to protect them on the way home. Do not use plastic bags – the fungus will sweat and quickly become a sticky mess.

Back in the kitchen the first temptation is to pick up a book on fungi and flick through the pages trying to find something that 'looks a bit like' the specimen in hand. While I must tell you that I have shamefully done this myself on occasions and have even seen experienced taxonomists, in desperation, do the same, it is not the way to go about it. Instead, leave your books on the shelf and do the right thing – study your specimens.

If you have absolutely no idea what something is (and with some groups of fungi, even if you do) the first step is to take a spore

print. This only really works with gilled fungi, but it is mostly these that cause serious and common forms of poisoning. Either cut off the cap (if you have specimens to spare) and lay it, gills down, on a sheet of white paper; or if you do not wish to damage your specimen make a hole in a piece of paper, gently push the stem through the hole and place the arrangement on an empty drinking glass. In either case cover the cap with a plastic lid or something similar to keep it moist. A reasonable spore print can be obtained in a couple of hours, though overnight is best. All being well you will see a spore print on the paper. Unfortunately it is not like a fingerprint – it is just the colour that you are interested in. This will range from white through innumerable shades of cream to yellow ochre, several shades of brown and on to black. There is also pink. Alone, the spore print colour will not give you a positive identification but it will eliminate a huge number of candidates. If the print is pure white then you have dispensed with a good 50% of possibilities. If it is pink then this rises to over 90%. Perhaps more importantly, it is spore colour that is often used as a starting point in the many 'keys' that have been written to aid identification (see below).

I always tell my botanist friends that they have things easy. Plants have so many highly variable visible characteristics that telling one plant from another is often simplicity itself. With mushrooms and toadstools, however, you have a cap and a stem and everything else is a matter of fine detail.

So what are these details? There is not room to list all of them here or to tell you in

what ways they might vary – any good book on field identification will tell you this. Briefly, however, it will include general size and shape, colour, whether it has gills, pores or spines, surface texture (scurfy, fixed scales, moveable scales, smooth, wrinkly, radiating fibres, dull, shiny, sticky, dry, etc.), presence and nature of a ring on the stem, presence of a bag (volva) at the base of the stem, peelability of the cap skin, and texture of the cap flesh and stem. There are many more. Each one, observed and noted, will provide a clue to the name of your find. There are many very difficult species which require specialist

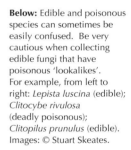

Right: St George's mushroom (*Calocybe gambosa*) is an excellent edible species with an aromatic flavour that appears in spring. Image: © John Wright.

a typical species from the various genera and a description of its defining characters. For example, there may be a picture of a horse mushroom as an example of the genus *Agaricus* and a note that members of this genus have a ring on the stem, brown gills which may be pink when young, a dark brown spore print, smooth white or brown scaly cap, can smell of mushroom, aniseed or sticking plasters, and that they are mostly medium to large in size.

Often there is a key which will tell you which genus it belongs to or even its precise name. Keys are notoriously difficult things to use (and far, far harder to write!). Conventional wisdom has it that they will nearly always provide an answer and it will nearly always be the wrong one! With experience, however, they can be very useful indeed. There are various types but the general idea is that you are asked a question which leads you on to another question, gradually narrowing down the field until a single genus or species has been decided upon. If the key tells you the genus then you will know which part of the book to look through. If your key gives a precise species you then check its picture and description in the book to determine if you are happy with what you have been told. In fact, check in as many books as you can find.

Very soon you will be able to recognise the more familiar genera such as *Agaricus* (e.g. the field and horse mushroom), *Russula* (brittlegills), *Boletus* (ceps, etc.), *Lactarius* (milk caps), *Amanita* (death cap and destroying angel *Amanita virosa* but also the delicious blusher *Amanita rubescens*) and *Macrolepiota* (the parasols). There are a couple of dozen of these larger genera and they are easily differentiated one from another.

A few final words on safety: I hardly need to remind you that some fungi are extremely poisonous, and some are quite deadly, so be absolutely certain that you know the common poisonous species. Of particular importance are the deadly death cap and the yellow stainer mushroom. The latter, though not seriously poisonous, will make you ill for a day or so and is by far the most common cause of poisoning. Last of all, never jump to conclusions – 'It looks a bit like the one that Carluccio was cooking with on telly last week' is not enough!

knowledge and the use of a microscope and other aids to identification, but these will seldom trouble the average mushroom hunter, whose prey will consist largely of distinct and well-defined species.

Having spent a few minutes observing all these characteristics and noting them down you will have come to know your fungus very well. Many books have a visual guide to point you in the right direction. They may have a picture of

Below: Saffron milk cap (*Lactarius deliciosus*). Image: © John Wright.

Where to forage

But where should one look for fungi and when? As the main part of a fungus is underground (or within a woody substratum) and usually perennial, many require long-lived habitats. Permanent grassland and mature woodland are typically good hunting grounds. But these need not be pristine wilderness. Road verges are permanent grassland and parks and churchyards are often mature woodland, at least as far as some fungi are concerned. The choice of woodland is very important, because of the mycorrhizal associations of many fungi as mentioned earlier. Beech, oak, birch, pine and spruce are the principal trees which support gourmet treats such as ceps, chanterelle, hedgehog mushrooms and the horn of plenty. There are a few edible woodland species that are not mycorrhizal, living instead on leaf litter or directly on dead or dying wood. Typical examples are the wood blewit (*Lepista nuda*), oyster mushroom (*Pleurotus ostreatus*), amethyst deceiver (*Laccaria amethystea*) and shaggy parasol (*Chlorophyllum rhacodes*). For a really large haul of mushrooms grassland affords superb opportunities. I have seen fields containing several hundred parasols, others with 50 horse mushrooms, a thousand field mushrooms, and, once, 26 giant puffballs (*Calvatia gigantea*).

When to forage

Autumn is always considered the main season for fungi and this is generally true. All my forays take place in late September and all through October. This usually works well but nature frequently confounds my plans and a wet summer followed by a dry autumn will effectively move the season forward by two months or more. Even in a 'normal' year there are some species that come early.

Above left: Wood blewit (*Lepista nuda*).
Image: © John Wright.

Above: Shaggy parasol (*Chlorophyllum rhacodes*).
Image: © John Wright.

Left: Giant puffball (*Calvatia gigantea*).
Image: © John Wright.

Giant puffballs, field mushrooms and chanterelle are frequently found in July and August. Late October and into winter will see wood blewits appear. The season for most species comes to a grinding halt with the first hard frost which kills off nearly everything not well protected. A mild spell after this may see some fungi return right up until Christmas or even beyond. Only two mushrooms of culinary interest are found in spring in Britain but they are among the best: the powerfully aromatic St George's mushroom (*Calocybe gambosa*) and the delicate and elusive morels.

Conservation

While the main worry for any mushroom hunter is whether or not his dinner is going to kill him, there are other matters to consider: conservation is one, the law of foraging is the other. Conservation of the fungi is a major issue (see Chapter 10), but it is a subject on which I feel much nonsense has been spoken and written.

Research in this area is extremely difficult but those studies that have been performed have provided no evidence that occasional picking of fungal fruit bodies has any effect on the long-term viability of a species. For anyone unfamiliar with how fungi grow this must seem very odd. If bee orchids were found to be a great gastronomic delicacy and a cottage industry was set up dedicated to pulling them out of the ground and sending them to Covent Garden then one would rightly be horrified at the inevitable decline of this attractive species.

However, collecting fungi is not equivalent to uprooting flowering plants. When one picks a mushroom one removes its reproductive organ – the main part of the organism, the mycelium, remains below ground, undisturbed (see Introduction). Though the parallel is not perfect, picking a mushroom is like picking an apple from a tree. But, one might say, surely the

fungus puts an enormous amount of effort into producing these fruit bodies and their removal must be detrimental to the long-term viability of the fungus. Well, it *may* be so but, I think, only if the species concerned is rare or the picking took place on a massive, long-term scale. One needs only to think of the humble bramble (which expends much energy in producing a huge crop of blackberries each year) to consider why this might be so – how many blackberries would one have to pick to affect the long-term viability of the bramble? The answer is 'nearly all of them' and for a long time. Fungi produce their spores in astronomical numbers (trillions) and eliminating a well-established species would be an extremely difficult enterprise (though see Chapter 10 for some of the very few case studies of fungi threatened by over-collecting).

People frequently tell me that the field where they picked mushrooms 20 years ago now produces nothing but dandelions. Invariably what has happened is that the field, once permanent grazing land, has been put under the plough and reseeded. The fungal mycelium has been destroyed. Where fungi have been lost in the UK it is nearly always as a result of habitat loss or pollution (see Chapter 10), not the efforts of the mushroom hunter.

Stories still abound of woodland, otherwise unchanged for decades, no longer producing the ceps or chanterelles they once did, with the finger of suspicion pointed at the mushroom hunter. These anecdotal accusations come from a failure to appreciate the fickle nature of fungi. They are not like most of the flowering plants which will flower every year. It may be that the mycelium is happily growing underground and that no fruit bodies will appear for many years – decades even. One of my cep spots, having been productive for five years, became barren for 12 years before resuming fruit body production for the last ten. Several fungi seem content to fruit very occasionally and, because they are so rarely seen, are assumed to be rare. It is not necessarily so. Some years they are found everywhere, showing us that they were probably there as mycelium all the time. In 2006, for example, a species I had long despaired of ever seeing – a splendid fungus called *Cortinarius violaceus* – found a year it was happy with and popped up everywhere.

So, does this mean we can pick whatever we like? No. Many species are genuinely rare and it would be unforgivable to reduce their chances further. On the forays I lead I allow restrained picking of common edible species and the collection of one or two specimens of less common species, edible or not, and rare species are observed *in situ*, noted, then left in peace. When collecting edible species for myself I seldom gather every fruit body from a location. I will leave the more mature specimens which are producing spores on an industrial scale and the young ones which have yet to produce any at all and pick just those in middle age.

There are other reasons for not picking everything. Many species of invertebrate make their living from mushrooms and toadstools, the fungus gnats (much hated as the maggots in everyone's cep à la crème) being particularly common. The wholesale removal of fruit bodies can seriously affect these organisms. Finally there is what town planners call 'visual amenity'. A wood that has been cleared of all its fungi has a sad, empty look about it – do leave whatever you can for others to enjoy seeing.

Left: Field mushroom (*Agaricus campestris*). Image: © John Wright.

The law

There is one more concern for the mushroom hunter – the legality of collecting fungi. It is all horribly complicated and not as settled as one would like. In common law in England and Wales, and enshrined in the 1968 Theft Act, there is a right to collect the four 'Fs' – fruit, flowers, fungi and foliage – provided that they are growing wild and that they are strictly for private use. As the Act says: he who does these things 'does not steal what he picks'. The two caveats, however, mean that it is not legal to pick flowers from a garden or apples from an orchard, though dandelions from a field or crab apples from a hedgerow are fine. If you sold your dandelion wine or crab apple jelly, however, this would be for commercial gain and the original collecting would be illegal. In Scotland the law is impenetrably vague and contradictory but picking for personal use is very unlikely to get you into trouble.

Scotland is not, however, plagued by the law of trespass and most Scottish land is accessible to all. In England and Wales there certainly is a law of trespass and you cannot go on to any land without permission. Fortunately it is not a matter for the criminal law and despite what the signs say you cannot be prosecuted for trespass, just asked to leave. Incidentally, picking the four 'Fs' while trespassing is not theft, it is a further act of trespass.

Matters have been made more complicated by the recent Countryside and Rights of Way (CRoW) Act 2000 – the so-called 'right to roam' legislation. This has given with one hand and taken away with the other. Collecting anything from land covered by the Act constitutes an act of trespass unless the right to collect existed before. If you are on land covered by the CRoW Act you cannot pick as much as a single blackberry. If you do you will be trespassing and must remove yourself immediately, not returning for 72 hours (perhaps to then pick another one).

Left: Bearded tooth, or lion's mane fungus (*Hericium erinaceus*). This basidiomycete is protected by law in the UK, and must not be picked. Though it is rare in nature it is easy to grow and can be cultivated for food. Image: © Martyn Ainsworth.

Many places, such as land owned by the Forestry Commission, National Trust or local government, provide open access and the picking of mushrooms may be allowed. Sometimes there is a daily limit per person. At the time of writing, for example, the Forestry Commission in the New Forest allows a generous 1.5 kg limit. Some land may have a ban on mushroom collecting and may have a by-law in place to enforce it. There is also conservation legislation to consider (see Chapter 10). Only four fungi are protected in law in the UK (Schedule 8, Wildlife and Countryside Act 1981). One of these, *Hericium erinaceus*, is edible and, surprisingly, is cultivated and sold dried in oriental emporia. It is illegal to disturb any of the four. On land which has attracted the eye of conservation bodies and been declared a Site of Special Scientific Interest (SSSI) or a Special Area of Conservation (SAC), some of the species which make it special will be 'cited'. Damaging such species is illegal. It is unusual for fungi to be cited but there is normally a 'catch-all' clause under the heading of 'activities likely to damage the site'. This invariably includes something along the lines of 'removal of or damage to any plant, animal, fungus or other organism'. Fortunately, the three national bodies responsible for the protection of these special sites, Natural England, the Countryside Council for Wales and Scottish Natural Heritage, tend to permit the sensible collecting of common plants and fungi.

Much has been made of the clause in the 1981 Wildlife and Countryside Act which prohibits the uprooting of any plant without the landowner's permission. It has been suggested that this clause applies to fungi – but, since mushrooms and toadstools are *not* plants, have no roots and are not even the whole organism, it is one I ignore.

I would ask you to enjoy your foraging experiences. Gathering food from the wild is, when one comes to think about it, the only natural way anyone can obtain food – all else is the product of human ingenuity. The foraging instinct is deep within us all and its expression a great and unexpected joy.

Calvatia gigantea (giant puffball) species profile

While the average mushroom hunter may appreciate the beauty and intricate design found in the fungal world, what they *really* want from a fungus is that it is tasty and, if possible, big. None fulfils this mycophagist's dream better than *Calvatia gigantea*, the giant puffball. It is a fungus I find two or three times every year, usually at my regular 'spots', and my heart still leaps every time I see one. I peel them and fry slices dipped in salted beaten egg to produce a fabulous wild omelette. There is always more puffball than I can eat so innocent visitors will often leave carrying half of one and looking a little bemused.

A good eating size for a giant puffball is 30 cm – about the same as a football. As they grow beyond this the dense flesh merely expands to become softer. They also need to be pure white, inside and out; if your find is showing any yellow/green then it has matured well beyond the stage when it can be eaten. I say 'matured' deliberately – it has not 'gone off' as many think.

When they produce their spores they do so in truly astronomical numbers. The frequently repeated figure of seven trillion from an average fruit body is perfectly reasonable. I once calculated that if the spores were the same size as a baked bean then seven trillion would fill the massive Millennium Stadium to the roof four times over. Puffballs are fairly common but not *that* common – why are we not knee-deep in white footballs? The answer is that most spores will land in unsuitable places and be lost, many will be eaten, and some will germinate and lose out in the battle for survival. But the fungal world is a secret one and many will have germinated to produce a viable mycelium but not yet produced fruit bodies.

John Wright

Image: © Sherry Stannard.

Main photos, clockwise from this page:

Brown puffball
(*Bovista nigrescens*).
Image: © RBGE/Robert Unwin.

Postia stiptica.
Image: © Peter Clarke.

Nectria cinnabarina.
Image: © Chris Jeffree.

Facing page inset photos:

Hygrocybe collucera (left);
Hygrocybe lanecovensis (right)
both from Australia.
Images: © Ray and
Elma Kearney.

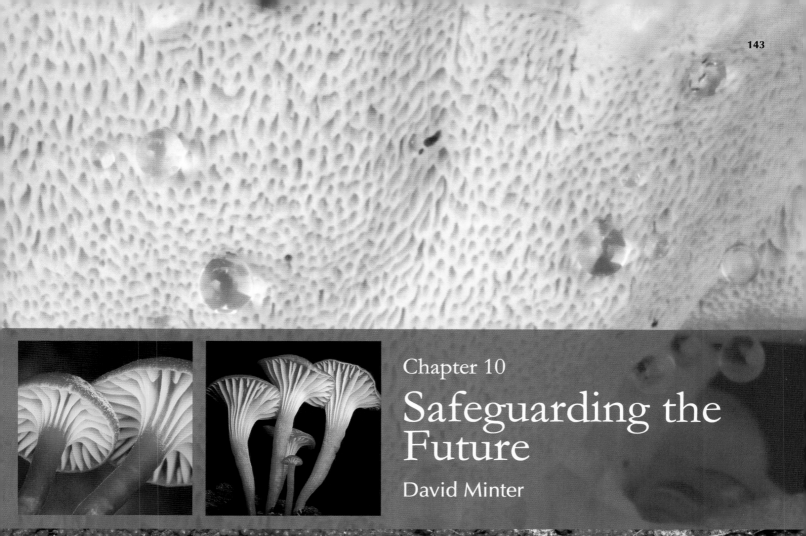

Chapter 10
Safeguarding the Future
David Minter

Safeguarding the Future

David Minter

There is now overwhelming evidence that human activity is causing the decline of animal and plant populations, with extinction predicted for many species. Fungi are just as vulnerable. Although remarkable in many ways, fungi have no special protection against humans. Mass extinctions of fungi can also be expected and, when they occur, many will be unrecorded, as so few of the world's fungal species are known (see Introduction and Chapter 1). Population decline in fungi was first noticed at least as far back as the early 1970s. Landmark work by Eef Arnolds demonstrated beyond any reasonable doubt that populations of larger fungi in the Netherlands were crashing. Since then, the evidence has been accumulating. This chapter looks at some examples of the environmental threats faced by fungi, and discusses what should now be done for fungal conservation.

Below: *Cyttaria espinosae* is one of the remarkable 'golf ball' fungi collected in Patagonia by Darwin during his voyage on HMS *Beagle* between 1831 and 1836. Its excellent flavour makes it a popular food, and basketfuls are a common sight in the market towns of southern Chile. Image: © David Minter.

Climate change is, almost literally, a hot topic. To date, the impact on fungi, their response, and the knock-on effect on humanity have only just begun to be considered. There is still little experimental evidence available, but mycologists have been analysing long-term records. There are already alarming indications that fruiting periods have changed for north temperate macrofungi (species with larger fruit bodies). Recent analysis of autumn fruiting patterns of larger fungi in Britain over the past half century showed that, for 315 species, the average first fruiting date is now earlier and the last fruiting date later. Many species are now fruiting twice a year, indicating increased mycelial activity and possibly greater decay rates in ecosystems. Changes are also being seen in the geographical distributions of fungi as they move in response to climate change; nobody knows whether these species on the move should be regarded as dangerous invaders or as refugees in need of help.

For almost all fungi, our current level of understanding about climate change impacts is pitifully low. In most cases we know nothing. The implications for agriculture and forestry, through potential spread of crop diseases, have been largely overlooked (see Chapter 3). Not enough is known about how fungi will respond to temperature, rainfall and carbon dioxide changes, nor about what impact that will have on, for example, the carbon cycle. Only a few fungi – those occupying habitats clearly threatened by global warming (alpine and arctic fungi, coastal and coral reef fungi, mangrove swamp fungi, etc.) – can be evaluated as under threat, even if only tentatively. Our example is the ascomycete *Lachnellula pini*.

Lachnellula species inhabit conifers and produce cup-shaped fruit bodies up to about 5 mm diameter on the surface of the bark.

When fully open, these cups reveal a beautiful bright orange disc surrounded by a fringe of hairs. In some species the hairs are white. In others they are brown. *Lachnellula pini*, one of the brown-haired species, occurs on Scots pine in Scandinavia. The species has been studied because, although beautiful to look at, it also causes a canker disease in the trees. That means there is at least some information about its distribution and biology. Research has shown that damage to trees caused by settling and creeping of snow provides a focus for infection, while its geographical distribution coincides with areas where the mean January temperature falls below a certain point. That research was done by foresters wanting to control a pest, but we can re-interpret the same information from a conservation standpoint: a species, restricted to the far north and mountainous areas with low January temperatures, and apparently dependent on snow damage for colonisation, may well be vulnerable to climate change.

Two important points are raised by this example. The first is that for fungi there is usually very little information available; the reader's response to the preceding paragraph is likely to be, 'Is that the best you can do?' Unfortunately, the answer is that, compared with most cup fungi, *Lachnellula pini* is well documented. As a result, statements about fungi and conservation have to rely heavily on inference. This is really not ideal. The second is that lists of endangered fungi are going to include species which have traditionally been viewed as harmful to our crops; conservationists need to be aware of the political and moral issues this is likely to raise.

Pollution takes many forms, some local in impact, others spread over huge areas. They include acid rain, nitrogen deposition, contamination with heavy metals, and radioactive pollution. Some activities which we may not regard as pollution – such as application of fertilisers – affect fungal populations as adversely as other types of pollution, and some of these activities – such as application of fungicides – knowingly and deliberately target fungi. As far as the fungi are concerned, they are all pollution.

The responses of fungi to these various types of pollution have been studied. There is a wealth of evidence to show that lichens are greatly affected by a wide variety of atmospheric pollutants. Some are more affected than others, and which species are present provides a good indication of the amount of sulphur dioxide pollution. Fungal populations in old, permanent pastures on soils poor in nutrients decline after fertiliser application. Ectomycorrhizal fungi (see Chapter 3) – fungi growing in association with the roots of forest trees – are similarly hit by acidification and nitrogen accumulation from air pollution. The same happens with ascomycetes associated with conifer leaf

Above: A Croatian poster highlighting edible fungi that are legally protected and should not be picked or disturbed in their natural habitat. Image: Kindly reproduced with permission of the Croatian Mycological Society.

litter: acid rain results in a sharp decline in populations of most species, and the sudden increase of a very small number of generalist species able to survive in such conditions.

Radioactive pollution brings its own highly specific problems. For example, species of the genus *Elaphomyces*, subterranean fungi known as 'false truffles', accumulate radioactive pollution more strongly than any other organisms. Since these fungi are an important component of the diet of wild boars, that pollution can make its way up the food chain, ending up on the plates of hunters. In some parts of Europe, hunting for wild boar is now discouraged for that reason. From all of these species affected by pollution, two examples are covered in more detail below: *Poronia punctata* (also known as the nail fungus), and *Erioderma pedicellatum* (the boreal felt lichen).

Poronia punctata is typically found on old dung of horses, ponies or donkeys and, exceptionally, of elephants in the northern hemisphere and montane tropics. Large enough to be identifiable in the field, it produces a circular disc-like structure on the surface of the dung, with numerous tiny fruit bodies embedded within the disc. The disc itself has a long stalk going deep into the substratum, giving the whole

fungal structure the appearance of a nail, hence its English name. The disc is white and the fruit body exit channels black, making the disc surface speckled. First recorded in 1753, it was not regarded as rare until the mid 1970s when a decline was noticed. Since then it has become one of very few ascomycetes worldwide to be recognised as needing conservation. The fungus is assessed as Vulnerable for the Netherlands, is listed as Regionally Extinct in the Red List for Finland, is on the Red Lists for Denmark, Estonia, Macedonia, Norway, Poland, Slovakia, Sweden, Switzerland and the UK, and has been considered for red listing in Belgium and listed as potentially rare in Croatia. The UK Biodiversity Action Plan for this species described it as 'possibly the rarest fungus in Europe'.

Fruit bodies only appear on older slowly decomposing dung, typically more than one year old. To survive in this habitat, the fungus must be able to out-compete other species which colonised the dung at an earlier stage. Not surprisingly, therefore, *Poronia punctata* has been found to be strongly antagonistic to other fungi, producing unusual antibiotics which limit the reproductive power of its competitors. This complex process of competition, however, seems also to have made the fungus very susceptible to the presence of non-natural products, such as additives, in the food of the animal producing the dung. There is a good deal of evidence that the fungus occurs only on dung of horses and ponies which have been fed unimproved pasture or hay. It has declined not only because the internal combustion

Below: The nail fungus (*Poronia punctata*) – so called because it looks like the head of a carpenter's nail – is typically found on horse dung. With the rise of motor cars and the decline of horse-drawn transport this fungus is now extinct in some areas and on the Red List of many European countries. Image: © Stuart Skeates.

engine has replaced the beasts of burden which produce its substratum, but also because the meadows where those remaining beasts graze have been 'improved' (by addition of fertilisers, pesticides and high yield grass varieties) and additives added to the feeds.

The important points raised by this example are that effects of pollution on fungi can be unintentional but far-reaching (who could imagine that additives to animal feed or fertilising of meadows would have such effects?) and that even apparently obscure microfungi may have their uses. Discovery of the interesting antibiotics in *Poronia punctata* came just at the time when the source fungus was being recognised as endangered – an eloquent example of how action to identify and protect rare and declining species of fungi can be beneficial to mankind.

Erioderma pedicellatum is one of only two lichens on the Red List of the International Union for Conservation of Nature (IUCN), where its status is Critically Endangered. Originally known from Norway, Sweden and Canada, this species is an epiphyte (i.e. grows on plants) with highly specific habitat requirements for old-growth oceanic conifer forests. It is now found only at one small site in Alaska and as two Canadian populations, in Newfoundland and Nova Scotia. The population has declined by 80% in recent years and is still declining, mainly because the species is highly sensitive to air pollution in the form of acid rain, but also through habitat loss

caused by logging which removes the trees on which it grows and affects the microclimate. The population in Nova Scotia is protected by the Canadian Federal Species at Risk Act, and is the focus of recovery efforts to protect the highly specific habitat and alert forest managers to the rare organism for which they are responsible. The important point raised by this example is that threats to fungi often do not come singly. This is a species for which various adverse factors are conspiring to drive populations down.

There are lots of reasons why people collect fungi. The commonest are for human food, for medicinal, recreational or religious purposes, and for cultural (ornamental or craftwork) use. In many parts of the world people collect fruit bodies of larger fungi for food. Frequently this is done by individuals for personal use and is unlikely to pose a threat (see Chapter 9), but commercial collection of wild fungi also occurs and both individual and commercial picking are on the rise. With globalisation, commercial picking is frequently carried out in low-income countries or countries with no native tradition of eating wild fungi, and the fruit bodies may be transported great distances fresh or dried to markets in high-income countries. It is

Above: The products of a fungus foray in Croatia. Foraging for fungi is a common activity in Croatia, but commercial collecting is well regulated. Commercial picking is only permitted for 32 species and collectors must be licensed. There are also regulations regarding the size and quantity of fruit bodies that can legally be collected. The list of fully protected species for which any picking is illegal runs to 314 species and includes some good edible mushrooms. Image: © Vladimír Jamnický.

Left: The critically endangered boreal felt lichen (*Erioderma pedicellatum*). Image: © Christoph Scheidegger.

Above: The oyster mushroom *Pleurotus nebrodensis* has a very restricted distribution limited to Sicily and is threatened by over-collecting as it is a desirable edible species. Image: © David Minter.

limits to picking, and uneasy on the other in the face of huge quantities of fungi – sometimes literally container-loads – being scooped up by commercial collectors, when common sense suggests that such levels must be having a harmful effect. Attempts to establish good practice codes and reasonable limits often meet unfavourable press coverage along the lines of 'now they even want to control us picking mushrooms'. It is also worrying that this international trade is not subject to constraints like the Convention on International Trade in Endangered Species (CITES), which explicitly covers only animals and plants. For a few fungi, however, there is good evidence that levels of collecting are unsustainable. Two examples are presented below.

Pleurotus nebrodensis is a species similar in general appearance to *Pleurotus ostreatus*, the well-known oyster mushroom. Known only from Sicily, it grows within an area of less than 100 km^2 on limestone mountain meadows in association with *Cachrys ferulacea*, a plant related to celery. This is currently the only mushroom on the IUCN Red List, where it is categorised as Critically Endangered. In recent years, road building and other developments have fragmented the population and, because the species is highly prized gastronomically, it is subject to acute pressure from collectors. It has no legal protection,

not difficult, for example, to find chanterelles collected in Russia or morels collected in Argentina being sold in substantial quantities in Spanish markets.

There are several studies of the impact on fungal populations of this sort of collecting. Most have tended to be inconclusive. The result has been that mycological societies and other nature organisations have been reluctant on the one hand to establish strong

Right and far right: Fruiting structures of *Ophiocordyceps sinensis*, a parasite of buried insect larvae found in the eastern Himalaya. Being prized for its supposed medicinal qualities this fungus is collected in enormous quantities each year. Image: © Paul Cannon (CABI/RBG Kew).

and efforts to prevent collection lack resources. As a result, it is estimated that each year fewer than 250 fruit bodies reach a spore-producing state. It is hard to see how any collecting can be sustainable under such conditions. Fortunately, like the oyster mushroom, this species is easily cultivated, and is now grown commercially in the hope that this will relieve pressure on wild populations.

The other example, *Ophiocordyceps sinensis*, is a small ascomycete which parasitises and kills insects (see Chapter 4). This species occurs in high-level meadows and associated bush ecosystems in the Himalaya and on the plateaux of Tibet. It has been prized in Chinese traditional medicine for hundreds of years as an aphrodisiac and as a treatment for various ailments. In recent years, price has increased with demand, reaching US$18,000 per kilogram in Tibet in 2008, and now the fungus is collected very extensively, with annual production perhaps exceeding 100 metric tons, an amazing figure given the tiny size of each fruit body. In Bhutan in the early 2000s, the value of the *Ophiocordyceps sinensis* harvest (mostly hidden in the black economy) was estimated as US$20 million per annum, a sum equivalent to the *entire* official value of all exports from Bhutan excluding energy. Given these enormous financial incentives, current collecting is at unsustainable levels, much of it beyond the law in the form of a fungal equivalent to poaching. A recent project funded through the UK Darwin Initiative has been helping authorities in Bhutan to develop conservation measures. Without these, the future of this fungus in the wild may be uncertain. Like *Pleurotus nebrodensis*, however, it too can be grown in culture, and there is now growing interest in commercial production of this species in cultivation.

The points from these two examples are that, where strong economic pressures exist, effective conservation of fungi in the wild may be difficult or even impossible, and that, where possible, *ex situ* conservation may sometimes alleviate these impacts.

The conservation of fungi also needs to take account of the fact that some species are found in only a small area or in highly specialised habitats. Some of those restricted species

may have had a wider distribution in the past, and their present limited range may be an indication that their preferred habitat has been destroyed, but others seem always to have had a restricted distribution. Where information is available, they can be the easiest to assess for conservation status, because the IUCN red-listing criteria set clear distributional limits for different levels of vulnerability.

An example of this is *Zeus olympius*, an ascomycete known only on twigs of a few trees of Bosnian pine from Mount Olympus in northern Greece, where it is apparently endemic: despite searches, it has not been found elsewhere or on other plants. Although possibly parasitic it seems to pose no serious threat to the pine. *Zeus olympius* produces circular fruit bodies below the bark and a blackened protective crust. When ripe and moist, the fruit body expands, breaking through the bark, and splitting the crust radially into small black teeth, thereby revealing the fertile layer – an orange- or golden-coloured disc about 5 mm diameter. Although part of a national park, the area with known colonies of this fungus is very small. The species may be particularly vulnerable as the known colonies are near picnic sites and adjacent to recreational trails. Forest fires, natural or

Above: *Zeus olympius*, a rare parasite of Bosnian pine found only in northern Greece. The fruit bodies are covered by a black crust (upper twig) that splits to reveal the orange or golden fertile layer (lower twig). Image: © David Minter.

otherwise, are becoming increasingly severe because of climate change, and are now a real threat. The conservation status, currently being assessed, seems likely to be Critically Endangered, given the extremely limited range of this fungus. With the extensive changes occurring in habitats as a result of human activity, species with a restricted distribution are often particularly vulnerable.

Because fungi, like animals but unlike plants, do not produce their own carbohydrates, a different type of restriction can occur when, for example, a parasitic fungus grows only on one particular species or genus of plant. In the world of fungi there are many specific associations of this sort, including examples where fungi are specific to a plant which is itself endangered. In such cases, it seems reasonable to regard the fungus as no less endangered than the plant on which it depends. A good example of this concerns fungi associated with *Coussapoa floccosa*. This is a tree which itself is red listed in Brazil and was thought to be extinct in the wild until rediscovery in three locations in the state of Minas Gerais. The combined population of those sites is no more than a dozen plants. A study of the fungi associated with living leaves of this tree turned up six ascomycete species previously unknown to science, one in a new genus. A search was made for these fungi on living leaves of adjacent plants of different species without success. Inoculation tests were then carried out using spores to try to infect leaves of *Coussapoa*

floccosa and other plants, and results indicated that all six would grow only on *Coussapoa floccosa*. Various conservationists have suggested that a significant proportion of the world's most endangered species are parasites, and that extinction rates based on plants and vertebrates alone, not taking these parasites into account, represent huge underestimates. The point from the present example – and it is a scary one – is that those conservationists are surely correct.

A curious situation is an endangered species never recorded from the wild. An example of this is *Diehliomyces microsporus*, a tiny truffle-like fungus, known only from compost used for commercial farming of mushrooms. Claims that it originates from soil, while plausible, lack supporting evidence: it seems never to have been found in any natural habitat. Infested compost typically smells of chlorine, while fruit bodies, which occur below the compost surface, are cream-coloured, more or less globose and up to about 3 cm diameter. Mushroom farmers regard *D. microsporus* as a 'fungus weed' because infestations can cause serious economic loss. In recent years various treatments have been developed, including compost pasteurisation. These very effectively control infestations, and in farms using the treatments, *D. microsporus* infestations have been totally eliminated. Because of the commercial advantage in having no infestations, the procedure is now widespread. The only habitat in which this fungus has ever been observed is thus being made unavailable to it. This species is endangered in cultivation and unknown in the wild – a very unusual situation and a classic example of how inadequate information makes realistic conservation status assessments impossible.

These few examples give some idea of the varied issues facing fungal conservation. Those issues can only be resolved if there is an infrastructure to handle them, much more information, and more people active in fungal conservation.

The fungal conservation movement is in its infancy, and public awareness of the importance of fungi is very low. This is reflected at governmental level by the fact that, in many countries, explicit legal protection for fungi

Below: *Hygrocybe collucera.* This beautiful wax cap is classified as an endangered fungus under New South Wales Threatened Species and Conservation Act. Image: © Ray and Elma Kearney.

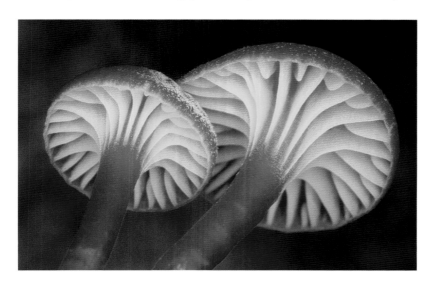

is totally absent. The national biodiversity action plans produced as a result of the Rio Convention are inadequate in respect of fungi, and some mycologists are describing fungi as 'the orphans of Rio'. Low public awareness also means big international conservation NGOs (non-governmental organisations), such as Conservation International, the World Wildlife Fund and the IUCN, fail to give fungi the prominence they merit, although this may be starting to change. Until early 2009, the IUCN included only two committees for fungal conservation in its Species Survival Commission, both listed under 'plants'. There are, in comparison, 20 committees just for birds. But the number of fungal expert committees has now risen to five, a step in the right direction but, given the diversity of fungi, hardly the end of the road. Mycologists are now working with the IUCN to deal with serious practical problems in applying their red-listing criteria to the fungi in the hope that more fungi will be formally evaluated through this system.

The broader conservation movement remains largely unaware of the need to conserve fungi. In Red Lists, frequently only one or two token larger fungi are listed as endangered, and then usually as 'lower plants', and fungi (e.g. host-specific species known only on rare endemic plants) may be treated as part of the problem (a threat to the plant) rather than recognised as being themselves in need of protection. Furthermore, the broader conservation movement usually identifies priority habitats on the basis of bird, mammal and flowering plant diversity, with fungi rarely taken into account. This can mean that habitats rich in fungal diversity are being missed by such categorisation.

There are very few NGOs specifically dedicated to fungal conservation, but encouragingly, the number is starting to rise. Mycologists in Europe co-operate through the European Council for Conservation of Fungi (ECCF), established in 1985 and now the conservation wing of the European Mycological Association. The ECCF is the oldest continental-level body devoted to fungal conservation, but other mycological

Left: 'Red Lists' identify threatened organisms based around criteria established by IUCN. Relatively few Red Lists of fungi have been published. At a European level there is only the *Red List of Macrolichens in the European Community*. This Red List is for fungi in Sweden.
Image: © RBGE/Max Coleman.

societies have also established conservation groups. Lichen conservation is covered by the International Association for Lichenology, while there are conservation committees for Africa, Asia, Australasia and North America, each run by the appropriate continental-level mycological society and mostly established since 2008. In South America, there is a specialist group for fungal conservation in Brazil, set up in 2007, and a steering committee was appointed in November 2008 to establish a conservation group for the Latin American Mycological Association: every inhabited continent thus has some sort of body devoted to fungal conservation. Finally, in October 2009, agreement was reached to establish a Global Federation of Fungal Conservation Groups. The challenge is now to do something with that basic structure for fungal conservation at last in place at an international level.

Efforts in fungal conservation to date have emphasised gathering scientific evidence for threats and decline (there have also been successful efforts in the UK to identify 'important fungal areas'). Evidence gathering has been directed mainly towards the visually more striking larger fungi, including species forming conspicuous lichens. Basic activities include compilation of distribution and ecological data into regional and national databases from which Red Lists are prepared.

National lists have been prepared for most European countries for basidiomycetes and some larger ascomycetes but coverage is patchy. For lichens, there is a preliminary *Red List of Macrolichens in the European Community* but, although efforts continue, there is still no European-level Red List for any other group of fungi. For other parts of the world, Red List coverage is usually poor to non-existent. There is clearly much more to be done, and if the work is to be extended to, for example, the tropics, then there is the problem of how to make relevant information accessible to the scientists in those countries:

most do not have library facilities comparable to those of Europe's or North America's great scientific institutions.

The information impediment is, in fact, enormous. There are various global initiatives to make biodiversity information available through the internet, but fungi have tended to be at the wrong end of the priority queue and mycologists have not been consulted at the design stage (the orphans of Rio, like orphans everywhere, are excluded from family decisions). As a result, these great internet biodiversity resources lack features essential for work with fungi: the flagship website for data on geographical distributions makes no provision for recording unsuccessful searches and, while its distributional records may be viewed through a fantastic mapping system, you cannot find out for any record what plant the fungus was growing on. More seriously, you cannot find out for any plant what fungi have been recorded growing on it. Whatever happened to the ecological approach to conservation?

Mycologists have therefore found it necessary to organise their own databases and websites. These function well, but – no surprises here – lack the levels of funding which global initiatives have attracted. Mycology thus suffers from the dreadful 'fauna and flora' approach, an intellectual straitjacket which

Right: *Hygrocybe lanecovensis.* This wax cap is classified as an endangered fungus under New South Wales Threatened Species and Conservation Act. Image: © Ray and Elma Kearney.

somehow makes otherwise rational scientists blind to the fungal kingdom: the oldest biological society in the world, the Linnean Society, still formally recognises only animals and plants. Until that approach has been discarded, those global initiatives are unlikely to adapt to cater for mycological information.

The 'fauna and flora' approach has resulted in mycology receiving miniscule and declining resources for biodiversity and conservation just when this work is more urgent than ever before. It has been said that 'to conserve fungi it is first necessary to conserve mycologists'. But in Britain, at present, mycologists with the traditional skills to identify fungi are more endangered than at any time in the whole of the 20th century and, ironically, the population decline steepened in the years following the Rio Convention. For mycology to be treated as a discipline of the same rank as botany or zoology, substantial human resources will be needed. Plans need to be put in place to train those people, and mycological societies need to be aware that their amateur component will be a key resource: these members are mycologists in their spare time, but professionally may well be the lawyers, journalists, press officers, web designers or data analysts who are so important for conservation.

The problems facing fungal conservation are daunting. They begin with science. It is no coincidence that this chapter contains no practical advice about how to conserve fungi. Evidence is all too frequently lacking or unavailable. The problem trail leads rapidly to informatics – the need for easy access to relevant data. Then there is the human resource question. Who will do the work? Who will provide the missing experimental evidence? But the difficulties do not end there. Mycologists need to recognise that conservation is not just science. It is also, and very much so, a political activity. That means lobbying. Scientists may perceive politics as a murky world, and they may not like going near it, but, to achieve conservation, it is unavoidable. This realisation has prompted the rapid development in the last three years of a basic infrastructure globally for fungal conservation. Now that infrastructure needs to be used to turn theory into practice.

Hericium erinaceus (bearded tooth) species profile

If fungi are allocated roles, this one is at the heart of the wood recycling team. In Britain, it specialises in occupying the central cores of beech trees and occasionally oaks, particularly those that have been open-grown in parks or wood-pasture. With such a large and secure food supply, it is not surprising that it can reside within a single tree for many years. Fruiting, however, is intermittent; some trees have been recorded producing fruit bodies on and off for two or more decades. It is always exciting, therefore, to visit one of these known trees to see if this rarity is on show. Its British stronghold is undoubtedly the New Forest, and to give an idea of what rarity means in this case, Alan Lucas and I could only find around a dozen New Forest 'fruiting beeches' during a dedicated survey.

Ted Green, who began visiting just such an occupied hollow beech in Windsor Forest in the 1960s, took me to see 'his' tree one October in the early 1990s with the warning that fruiting could never be guaranteed. I had never seen the fungus and so, with expectations suitably lowered, I was totally fascinated as the white 'football' adhering to the side of a distant trunk resolved itself into a stunning structure resembling a sculpture of a frozen waterfall.

The spores are dropped from slender, white, spaghetti-like spines or teeth (hence the English name) which can reach several centimetres in length. Indeed most of the fruit body consists of this beard-like mass of spines although a few short stubby branches can be found near the point of attachment. Several books list bearded tooth (also known as lion's mane fungus) as edible. For those wishing to investigate whether it really tastes like lobster (as some claim), I would point out that this species is protected by law. However, this fungus is not completely off the menu since it is easy to cultivate and is commercially available. The puzzle is why does it fruit so easily in cultivation but is so rarely seen in the wild?

Martyn Ainsworth

Image: © Martyn Ainsworth.

Main photos, clockwise from this page:

Pickled mushrooms; mushroom pancakes; honey fungus with pasta. Images: © RBGE/Sadie Barber and Vlasta Jamnický.

Facing page inset photos:

Warm salad of wild mushrooms (left). Image: © Martin Wishart Limited 2010.

Thick and chunky mushroom soup (right). Image: © Richard Milne.

Recipes

Ljerka Jamnický, Jevgenia Milne,
Andy Overall and Martin Wishart

Cautionary notes:
- be certain mushrooms are not poisonous
- consult an expert if unsure
- if in doubt don't eat
- the first time you try a new mushroom only eat a little

Pickled mushrooms

by Jevgenia Milne

This recipe and the following one have been used by my family in Estonia for three generations. Similar recipes will no doubt feature in many Eastern European cookbooks, but the following ones have been adapted and adjusted to our particular taste, their precise origin long since forgotten. The quantities are not very exact – it is often a case of adding ingredients by feel rather than by the measuring jug – but I've tried to make the basic guidelines as clear as I could.

Pickled mushrooms make a great antipasti and can be served straight out of a jar. Most edible mushrooms taste great pickled. Small mushrooms (up to 4 cm) cooked whole look especially appealing; but larger specimens can also be used cut in half or quarters.

Ingredients

Quantities to suit

Mushrooms

Water

Salt

Vinegar
(apple cider or wine vinegar)

Optional extras to spice up
the marinade may include:

Black pepper

Pimento (allspice)

Cloves

Bay leaf

Sugar

Image: © RBGE/Sadie Barber and Vlasta Jamnický.

Method

1. Clean the mushrooms of any forest debris and damage, and peel those that easily lend themselves to it. Washing is unnecessary, since parboiling will help get rid of any remaining dirt.

2. Parboil the mushrooms in plenty of water and drain in a colander, discarding the water.

3. Meanwhile, prepare jars that can be hermetically sealed – the screw-top large jam jars or similar are ideal. Wash and sterilise the jars in a microwave or in the oven, or using boiling water to dip the tops and then pouring the boiling water into the jars (up to 1/3 volume), covering them with clean newspaper to let them steam up, and discarding the water once they are thoroughly hot. Sterilise the lids in boiling water as well – they can be left in the water until needed.

4. Mix the marinade ingredients in a large pot. Water, salt and vinegar are the essential ingredients, others can be added to taste (black pepper, pimento, clove, sugar etc.). The amount of water must be sufficient to cover the mushrooms, and other ingredients are added to taste. The marinade must taste far too salty and sour!

5. Bring the marinade to the boil, and add the drained mushrooms. Boil for a further 5–10 minutes.

6. Lift the mushrooms out into the prepared jars and top with enough marinade to cover them. The remaining marinade can be reused on another batch of mushrooms.

7. Put the lids on while the jars are still very hot, using tongs to lift the lids out of the hot water and an oven glove to screw them on. Leave to cool.

8. Place the jars in a cool place to keep. If sealed hermetically, they should keep for months.

Thick and chunky mushroom soup

Russian-style
by Jevgenia Milne

This soup can be cooked with fresh mushrooms, fried mushrooms, or rehydrated dried mushrooms. It is very good made with ceps (*Boletus edulis*) or other edible boletes, but also a great way of using two-toned scale heads (*Kuehneromyces mutabilis*), or honey fungus (*Armillaria mellea*) – for those not allergic to it. For an authentic experience, serve with a dollop of sour cream and a slice of black rye bread.

Method

1. Fill a pot (approximately 3 l capacity) with water, add the drained barley and bring to the boil.

2. Meanwhile, if using fresh mushrooms, clean and chop thinly.

3. Once the soup water is boiling, add the potato pieces.

4. In a frying pan, sauté the carrots until they begin to soften and add to the soup pot. Repeat with the onions.

5. Bring the soup to the boil, then add the chopped mushrooms and herbs.

   If desired, you can also sauté the fresh mushrooms before adding them to the pot – some argue that this brings out their flavour better. However, two-toned scale heads and honey fungus are best boiled directly.

6. Season and leave to simmer for 5–10 minutes or until the barley, potatoes and carrots are all cooked.

Options

1. If using dried mushrooms, pre-soak in cold water for 30 minutes, chop and add to the pot after the barley and before the carrot and onion.

2. If using lighter-coloured and -textured mushrooms such as two-toned scale heads, adding some mushroom stock will enhance the flavour of the soup.

3. If you are out of pearl barley, or want to speed up the cooking, use some fast-cooking oat flakes instead, adding them after the carrots and onions.

Ingredients

Serves 6

Mushrooms, preferably ones that can be cooked straight (i.e. do not require parboiling – if using these, parboil and drain, discarding the water). You'll need a large handful of fresh or a small handful of dried mushrooms.

Approximately 75 g pearl barley, soaked overnight

2 medium potatoes, peeled and chopped into small cubes

Approximately 2 l water

1 carrot, peeled and sliced

1 onion, peeled and chopped

Herbs to taste (dill, bay, parsley), fresh or dried

A little oil for frying

Seasoning to taste

Image: © Richard Milne.

Mushroom pancakes

by Ljerka Jamnický

A favourite leisure activity in Croatia is hillwalking. Picking wild food along the way is one of the great pleasures of being outdoors.

Growing up in Zagreb we were fortunate in having the densely forested Mount Medvednica massif on our doorstep. I have fond childhood memories of Sundays on the mountain, returning in the afternoon with bowlfuls of wild strawberries and a selection of fungi. Dinner could then be a delicious one: a thick wild mushroom and potato soup, followed by strawberry-filled pancakes.

These days, if we have friends for a meal, I often make pancakes with mushroom filling. Fresh ceps are ideal on their own, but the common cultivated mushrooms combined with dried ceps for flavour are a good stand-in.

Ingredients

Makes about 16 pancakes

Batter

3 eggs

160 g white flour

300 ml milk

110 ml water

Pinch of salt

Oil for cooking pancakes

Filling

2 tablespoons cooking oil

2 onions

25 g dried ceps (*Boletus edulis*)

750 g fresh mushrooms (wild or cultivated)

Salt

Black pepper

Parsley

4 tablespoons double cream

Oil or melted butter to brush on top

Grated Parmesan (optional)

Method

1. Finely chop and sauté the onions in oil.

2. Finely chop fresh mushrooms, crush dried ceps with your fingers and add all this to the onions once they are soft. Season to taste.

3. The mushrooms will release juice, so continue simmering, stirring periodically, until all the liquid evaporates.

4. Leave to cool.

5. Mix in ground pepper, chopped parsley and cream.

6. Make pancake batter in a blender.

7. Leave to stand for 30 minutes.

8. Try to make one pancake. If it turns out too thick, add more water to the batter and cook the rest, keeping the batter stirred as you go.

9. Fill the pancakes. Just roll them up – or if you like, tuck the ends in.

10. Arrange in an ovenproof dish in one or two layers, and brush over the top with oil or melted butter.

11. Sprinkle with Parmesan, if you wish.

12. Bake in a preheated oven, 30 minutes at 190 °C or until crisp on top.

13. Serve with green salad.

Image: © RBGE/Sadie Barber and Vlasta Jamnický.

Honey fungus with pasta

by Ljerka Jamnický

More often than not we return from a foray with a number of species but just a few specimens of each. A mixed mushroom soup is then an obvious choice, but if you have a worthwhile quantity of a single species there can be other options – for example, *Lactarius deliciosus* just fried in butter, or puffballs sliced, breaded and fried. If we come across an old tree stump covered by honey fungus (*Armillaria* species) there is no dilemma – they are delicious cooked with pasta.

This recipe is a variation on one from south-western Croatia. Pasta would originally be home made, but supermarket pappardelle, fettuccine or tagliatelle make a good substitute. If honey fungus is not available cultivated shiitake mushrooms will be just as good.

Method

1. Clean the mushrooms by brushing off any soil, etc. The spores of honey fungus are white, and on a mature clump the lower caps are often dusted white by spores from those above; this superficially looks as if the caps are mouldy. Young specimens can be used whole, but discard tough stalks of mature ones and use just the caps.

2. Coarsely chop the mushrooms, garlic and parsley. Sauté the mushrooms and garlic with a pinch of salt in a half–half mix of butter and olive oil, stirring until the liquid evaporates. Add cream mixed with lemon juice, add the parsley and simmer for a few minutes.

3. Meanwhile cook the pasta.

4. Mix with the mushrooms; add ground pepper and salt to taste, sprinkle with grated Parmesan and serve.

Cautionary note: *Armillaria* can cause gastrointestinal upsets in susceptible people. Some sources recommend parboiling and draining before use, but we have not found this necessary.

Ingredients

Serves 4 to 5

400 g dry pasta (or 600 g fresh)

500 g mushroom

Olive oil

Butter

Parsley

Clove of garlic

150 ml single or double cream

10 ml fresh lemon juice (2 teaspoons)

Parmesan

Ground black pepper

Salt

Image: © RBGE/Sadie Barber and Vlasta Jamnický.

Oven baked *Hygrophorus marzuolus*
by Andy Overall

Hygrophorus marzuolus is a highly prized edible mushroom, found in southern Europe and north Africa. It can sometimes be bought from natural produce stalls in Spain, though it is often somewhat elusive for purchase and in nature. It fruits during March to June at the latest. On a mission to photograph and record the species in the foothills of the Sierra Guadarrama some 50 km north of Madrid, Spain, we discovered masses amongst the huge mature Scots pine. It makes a delicious starter to a meal. A real treat.

The southern distribution of this mushroom means that you may be unlikely to encounter it. In addition, *Hygrophorus marzuolus* is protected in certain countries. If you are not fortunate enough to be able to use *Hygrophorus marzuolus* good alternatives to use in this recipe are St George's mushroom (*Calocybe gambosa*), chanterelle (*Cantharellus cibarius*), wood blewit (*Lepista nuda*) or field blewit (*Lepista saeva*).

Ingredients

Quantities to suit

Mushrooms

Olive oil

Garlic

Prawns

Pepper

Salt

Method

1. The mushrooms are cleaned and kept whole, then brushed with olive oil and crushed garlic.

2. Sprinkle with salt and pepper.

3. Place in a preheated oven at 150 °C, gills up.

4. Once cooked garnish the mushrooms with sauteed prawns.

Note: These mushrooms are known to go well with seafood.

Image: © Andy Overall.

Warm salad of wild mushrooms

with watercress and balsamic vinegar by Martin Wishart

I often use mushrooms in my recipes and there is such a fantastic selection of varieties, sizes and flavours available.

For this dish I have used chanterelles (*Cantharellus cibarius*) and hedgehog mushrooms (*Hydnum repandum*). If you don't pick your own these mushrooms can be sourced from your local farmers' market or delicatessen when in season.

Scotland has wonderful wild mushrooms, with an abundance of choice edible species. I buy my mushrooms throughout the year from a number of mushroom suppliers across Scotland and France.

Three varieties that I use regularly in the restaurant are chanterelles which have a delicious autumnal earthy taste, ceps (*Boletus edulis*), which are one of my favourite mushrooms and have a meaty taste and texture, and the horn of plenty (*Craterellus cornucopioides*).

Method

1. Prepare the mushrooms by removing any tough stalks and cutting into evenly sized strips.

2. Wash the prepared mushrooms quickly in a large basin of water and drain. Leave to air dry on a wire rack for 30 minutes.

3. Peel one of the garlic cloves and cut it in half. Add it to a pan with 2 tablespoons of the olive oil and heat it to infuse.

4. Cut the bread into 1cm squares and fry these in the olive oil on a medium heat until golden brown. Drain through a colander or sieve and dry on a paper towel.

5. Lightly season the croutons with salt and remove the garlic.

6. Put the remaining oil into a wide-based pan.

7. Peel and dice the shallots finely and fry in the oil for 30 seconds.

8. Add the remaining garlic, peeled and halved, to the pan and then add the mushrooms in one even layer.

9. Season with a little salt and sauté until golden brown. Transfer to a mixing bowl.

10. Add the balsamic vinegar to the mushrooms while still hot.

11. Remove the garlic and allow them to cool a little.

12. Remove and discard the watercress stalks and rinse the leaves in cold water.

13. Toss the leaves into the lukewarm mushrooms and arrange on a plate.

14. Top the salad with the croutons and serve.

Ingredients

Serves 6

1 kg mixed wild mushrooms e.g. oyster, chanterelle, hedgehog mushroom

2 slices day old white bread, crusts removed

2 cloves garlic

2 shallots

5 tablespoons olive oil

1 tablespoons balsamic vinegar

250 g fresh watercress

Pepper

Salt

Lepista flaccida.
Image: © Peter Clarke.

Bibliography

Arora, D. (1986). *Mushrooms Demystified*. Berkeley, USA: Ten Speed Press.

Buczacki, S. (2011). *Collins Fungi Guide*. London: HarperCollins.

Cooper, M. R., & Johnson, A. W. (2003). *Poisonous Plants and Fungi* (second edition). London: HMSO.

Dobson, F. S. (2005). *Lichens: An Illustrated Guide to the British and Irish Species* (fifth edition). Slough: The Richmond Publishing Co. Ltd.

Findlay, W. P. K. (1982). *Fungi: Folklore, Fiction and Fact*. Slough: The Richmond publishing Co. Ltd.

Fox, R. T. (ed.) (1999). *Biology and Control of Honey Fungus*. Andover: Intercept.

Halstead, A., & Greenwood, P. (2009). *RHS Pests and Diseases*. London: Dorling Kindersley.

Hobbs, C. (1995). *Medicinal Mushroom: An Exploration of Tradition, Healing and Culture*. Santa Cruz, USA: Botanical Press.

Ingram, D. S., & Robertson, N. (1999). *Plant Disease, a Natural History* (Collins New Naturalist). London: HarperCollins.

May, W. J. (1898). *Mushroom Culture for Amateurs*. London: The Bazaar, Exchange and Mart.

Money, N. P. (2002). *Mr. Bloomfield's Orchard: The Mysterious World of Mushrooms, Molds and Mycologists*. New York, USA: Oxford University Press.

Money, N. P. (2004). *Carpet Monsters and Killer Spores: A Natural History of Toxic Mold*. New York, USA: Oxford University Press.

Money, N. P. (2006). *Triumph of the Fungi: A Rotten History*. New York, USA: Oxford University Press.

Moore, D. (2001). *Slayers, Saviors, Servants and Sex*. London: Springer.

Moore, D., Nauta, M. M., Evans, S. E., & Rotheroe, M. (eds.) (2001). *Fungal Conservation: Issues and Solutions*. Cambridge: Cambridge University Press.

Phillips, R. (2006). *Mushrooms*. London: Macmillan.

Purvis, W. (2000). *Lichens*. London: The Natural History Museum.

Ramsbottom, J. (1953). *Mushrooms and Toadstools: a Study of Activities of Fungi* (Collins New Naturalist). London: HarperCollins.

Spooner, B., & Roberts, P. (2005). *Fungi* (Collins New Naturalist). London: HarperCollins.

Stamets, P. (1993). *Growing Gourmet and Medicinal Mushrooms*. Berkeley, USA: Ten Speed Press.

Stamets, P. (2005). *Mycelium Running: How Mushrooms Can Help Save the World*. Berkeley, USA: Ten Speed Press.

Sterry, P., & Hughes, B. (2009). *Collins Complete British Mushrooms and Toadstools: The Essential Photographic Guide to Britain's Fungi* (Collins Complete Guides). London: HarperCollins.

Watling, R. (2003). *Fungi*. London: The Natural History Museum.

Wright, J. (2007). *Mushrooms: River Cottage Handbook No. 1*. London: Bloomsbury.

Useful websites

Ann Miller's Speciality Mushrooms
www.annforfungi.co.uk

The British Lichen Society
www.thebls.org.uk

British Lichens
www.britishlichens.co.uk

British Mycological Society
www.britmycolsoc.org.uk

MykoWeb
www.mykoweb.com

Fungi Online
www.fungionline.org.uk

Fungi Perfecti
www.fungi.com

OPAL Air Survey
www.opalexplorenature.org/?q=AirSurvey

Scottish Fungi
http://sites.google.com/site/scottishfungi

Tom Volk's Fungi
botit.botany.wisc.edu/toms_fungi/

The University of Edinburgh: Fungal Cell Biology Group
129.215.156.68/index.html

University of East Anglia: Fascinating Fungi
www.uea.ac.uk/bio/joyoffungi

Biographies

Lynne Boddy

Lynne Boddy is professor of mycology at Cardiff University, where she teaches and researches into fungal ecology, a topic that has fascinated her for over 35 years. She is (2009–2010) president of the British Mycological Society and chief editor of *Fungal Ecology*. Lynne is an ardent communicator of the mysteries and importance of the amazing hidden Kingdom of Fungi to the general public having, for example, taken part in a public debate on 'what are the most important organisms on the planet?' – speaking for fungi, of course. She jointly put together a gold medal winning display on the role of fungi, in the continuing education section at the world famous RHS Chelsea Flower Show.

Paul Dyer

Paul Dyer is a reader in fungal biology at the University of Nottingham, UK. He graduated with a PhD in mycology from the University of Cambridge, having become fascinated with fungi during his undergraduate studies at Cambridge. He has particular research interests in fungal sexual reproduction and fungal genetic variation. Study organisms include *Aspergillus* and *Penicillium* species, plant pathogens and lichen-forming fungi, and Paul has worked in Antarctica to study the population biology of Antarctic lichens. He is the current chair of the British Mycological Society Education and Outreach Group, which promotes fungal biology to schools, universities and the general public.

Harry C. Evans

Harry Evans has been involved with tropical mycology since completing his PhD in 1969 and taking up a UK government-sponsored position as cocoa pathologist in Ghana. He continued working on cocoa diseases in Ecuador and Brazil, as well as fungal diseases of endemic pines in Central America, before returning to the UK to take up a position with CAB International as plant/insect pathologist with particular reference to the biological control of invasive species. Harry developed an interest in entomopathogenic fungi during his time in the tropics. He is now a CABI Emeritus Fellow based at the UK-Egham Centre.

Stephan Helfer

Stephan Helfer was born in Southern Germany. He gained his first degree at the University of Tübingen, where he specialised in a group of fungal pathogens called the rusts. After post-doctoral work in Aberdeen, Stephan took up a post as electron microscopist/mycologist at the Royal Botanic Garden Edinburgh. This developed into full-time mycology with a remit to investigate pathogenic micro-fungi, such as rusts, mildews, smuts and wilt pathogens such as the Dutch elm disease fungus. His current work is split between research on European rusts and advisory and research work on pathogens of garden ornamentals.

Patrick Hickey

Patrick grew up in Scotland and graduated in plant science at the University of Edinburgh in 1997. He went on to complete a PhD on the cell biology of fungi. His research interests include studying the biology of hyphal tip growth, tracking organelle transport within fungal colonies and cultivation of bioluminescent and edible fungi. Patrick has developed novel imaging techniques for analysing the living cells of fungi including time-lapse microscopy of fungal mycelium labelled with fluorescent proteins and dyes. Patrick currently runs a technology company, 'NIPHT Limited', which designs and builds intelligent lighting systems based on LEDs.

Heather Kiernan

Heather Kiernan, a freelance writer and editor of intellectual and cultural history, was educated at the universities of Toronto and Cambridge. The recipient of The Scottish Film Council Award for her documentary film treatment *Chaplin — A Comedian Sees the World*, Heather went on to edit the memoirs of Georgia Hale, *Charlie Chaplin: Intimate Close-Ups*, and was commissioned to write a monograph on Chaplin for a series on American Intellectuals. She has just completed co-editing *Woodlanders: Life Among the Trees* (October 2010), a book celebrating woodland culture.

Naresh Magan

Naresh Magan has worked at Cranfield University for over 20 years where he has a personal chair in applied mycology. He has had a fascination for fungi since his BSc and MSc in botany and plant pathology at Exeter University in the 1970s. The work of his group has been focussed on the practical application of fungal technology and prevention of economic losses which can be caused by fungi. The use of fungi in enhancing breakdown of chemical contaminants in soil ecosystems, the potential for using volatile fingerprints as biomarkers and the production of pharmaceutically useful compounds have been other areas of interest.

David Minter

Inspired at school by Fraser Darling, James Fisher and George Waterston, those great field biologists who placed their mark on post-war nature conservation, David Minter has always been interested in observing and recording the natural world. Like so many, he began with an interest in birds, gradually transferring his allegiance to smaller and smaller organisms. Becoming a mycologist therefore by chance rather than design, he has worked with fungi all his professional life, and over the last 20 years has concentrated on developing computerised information resources for mycology, as tools to promote an appreciation of the fungi and work on their protection. Most recently he has established the website www.cybertruffle.org.uk as a means of making those resources freely available.

Nick D. Read

Nick Read has been a professional fungal biologist for over 25 years. He was inspired by the extraordinary world of fungi whilst doing his undergraduate degree in microbiology at the University of Bristol in the 1970s. In 1985 he took up a position of lecturer in mycology at the University of Edinburgh and has been there ever since. He is now a professor of fungal cell biology at the University. During his career he has published over 100 papers and has been invited to present over 100 talks on various aspects of fungal biology. He has played a very active role in the British Mycological Society (including being vice president and publications officer) and was the chair of the Organizing Committee for the 9th International Congress (IMC9: the Biology of Fungi) held in Edinburgh in August 2010.

Geoff Robson

Geoff Robson developed his interest in fungi during his first degree in botany and biochemistry and obtained his PhD from the University of Manchester where he is currently a senior lecturer. He has worked for over 25 years on various industrial, environmental and clinical aspects of filamentous (mould) fungi, published numerous scientific papers in these areas and supervised over 30 PhD students. Geoff is the outgoing secretary general of the International Mycological Association and previously was general secretary of the British Mycological Society. He is currently a senior editor of *Fungal Biology Reviews*.

Gordon Rutter

Gordon Rutter has a degree in fungal genetics and an MSc in plant taxonomy and biodiversity specialising in, of course, fungi. Fungal interests have taken Gordon all over the world including North and South America and extensive travels in Europe. Research in fungi at the Royal Botanic Garden Edinburgh and serving on the British Mycological Society's Education group and Council have furthered Gordon's interests and knowledge of fungi. As well as an interest in mycology Gordon is also fascinated by all aspects of the paranormal although he doesn't necessarily believe in it. The two interests coincide nicely with fungal folklore. A further sideline in photography complements these two interests. Gordon is currently based in Edinburgh working as a teacher of biology with a sideline in paranormal books, articles and lectures.

Milton Wainwright

Milton Wainwright is a lecturer and researcher in the Department of Molecular Biology and Biotechnology, University of Sheffield and Honorary Professor and Honorary Fellow of the Centre for Astrobiology, Cardiff University. He has wide interests in microbiology including recent ongoing research on the isolation of bacteria and fungi from the stratosphere, work which relates to his interest in panspermia, i.e. the theory that life on Earth originally came from space and continues to do so. He also actively researches the history of the germ theory and the development of antibiotics, notably penicillin. He has written three books, *Miracle Cure: the Story of Antibiotics*, *An Introduction to Fungal Biotechnology* and *An Introduction to Environmental Biotechnology*.

Roy Watling

Roy Watling, until retirement, was Senior Principal Scientific Officer and Head of Mycology and Plant Pathology at the Royal Botanic Garden Edinburgh. He graduated with a BSc from the University of Sheffield before gaining a PhD from the University of Edinburgh and DSc from his former alma mater. He was awarded the MBE by Her Majesty the Queen in 1996 and is a Fellow of the Royal Society of Edinburgh, from whom he received the Patrick Neill Medal for Excellence in Natural Sciences. He has been a committee member or fellow of many mycological and learned societies. He has maintained links with the mycological work of the Royal Botanic Garden Edinburgh and continues to communicate his passion for fungi.

John Wright

John Wright's interest in the fungi stretches back over 40 years and in 1992 he began teaching field mycology at adult education classes. He has led over 300 fungus forays and now runs a couple of dozen every year all over the country as well as many on other aspects of wild food. He writes on the subject of the edible fungi and wild food in general for magazines and newspapers and has written three books under the 'River Cottage' banner: *Mushrooms* (2007) – shortlisted for the Guild of Food Writers' 'Best First Book' award, *Edible Seashore* (2009) and *Hedgerow* (2010). He also spreads the word with television and radio appearances.

Glossary

Acaricide – A pesticide that kills members of the Arachnida, which includes ticks, mites and spiders.

Alga (plural algae) – Organisms that can make sugars by photosynthesis. They range in size from single cells to giant kelp.

Anaerobic – An anaerobic organism grows in the absence of oxygen gas. Most fungi grow aerobically; some prefer to grow in oxygen, but can grow in the absence of oxygen if necessary. One group of fungi in the guts of animals grows only without oxygen, so that they are obligate anaerobes.

Arbuscular mycorrhizas – An intimate association between fungi in the group *Glomeromycota* and the roots of plants (especially herbaceous plants). Fungi penetrate into the root cells of the host plant and very extensively in soil. The fungi feed water and nutrients to the plant and the plant provides the fungus with sugars.

Ascomycetes – These are commonly known as the ascus-producing fungi or sac-forming fungi. They produce their sexual spores in sacs, each commonly containing eight spores.

Ascomycota – See ascomycetes.

Ascospores – These are the sexually produced spores of an ascomycete. They are produced in an ascus (sac).

Ascus (Plural asci) – A sac-like structure in which ascospores (typically eight) are produced.

Bacterium – A microscopic single-celled organism, that can be shaped as a sphere, rod or spiral.

Ballistospores – A spore which is violently discharged to a short distance from the cell that produced it.

Basidiomycetes – The group of fungi that produce sexual spores in structures called a basidium. The basidium cells are formed, for example, in the gills of toadstools. The toadstools that we commonly see are basidiomycetes.

Basidiomycota – See basidiomycetes.

Binary fission – When a cell, e.g. yeast, reproduces by dividing into two.

Biomass – The weight of living organisms.

Biomonitor – Organisms that provide information on the quality of the environment. For example, lichens with different sensitivities can be used to monitor pollution such as by sulphur dioxide.

Biopesticides – These have the ability to kill invertebrate pests but are not chemicals – they are other organisms, very often microorganisms including fungi.

Biotrophy – The ability to feed from living cells of other organisms.

Budding – Many yeasts can reproduce by a small 'daughter' cell forming as a projection on the side of the large 'mother' cell. A cell wall completely surrounds the daughter so that it functions separately from the mother though they may still remain joined.

Chytrid – The short name for fungi in the group *Chytridiomycota*. These fungi can spread by the production of spores that can swim. They do not produce very much mycelium, but are able to colonise the cells of plant and animal material. Some colonise dead tissues whereas others can cause diseases of plants and animals.

Chytridiomycota – See chytrid.

Conidium (plural conidia) – An asexual spore that is produced at the end of a hypha, often in chains, and that is unable to swim.

Coprophiles – Organisms that grow on dung.

Cords – See mycelial cord.

Crustose lichens – Lichens whose thalli (bodies) form a crust on surfaces of, for example, rock or tree bark.

Cyanobacteria – Also known as blue green-algae, these are a group of bacteria that obtain their energy from photosynthesis and can obtain nitrogen from air.

Deuteromycetes – This group of fungi no longer exists. Originally fungi were placed in this grouping if they did not reproduce sexually.

Ectomycorrhizas – A mutually beneficial association between fungi and tree roots in which the fungal hyphae (filaments) sheath the young roots, grow between (but not within) root cells, and extend into the soil.

Endophyte – A fungus that is found within living plant cells.

Entomopathogens – Fungi that kill insects.

Entomophthorales – A group of fungi that kill insects.

Enzymes – Biological molecules (certain proteins) that break big molecules into smaller ones.

Exoskeletons – Vertebrates have skeletons inside their bodies. Invertebrates have their skeleton on the outside. For example, the hard external layer of a beetle forms its exoskeleton.

Foliose lichen – Lichens whose bodies are leaf-like in shape.

Fruticose lichen – Lichens whose bodies are 'shrubby' in shape.

Glomeromycota – A group of fungi that are intimately associated with the roots of plants in a mutualistic way (forming arbuscular mycorrhizas).

Herbivore – An organism that feeds mainly on living plants.

Heterotroph – An organism that is unable to perform photosynthesis but gets its energy from organic compounds that have already been made by other organisms.

Hypha (plural hyphae) – The basic structural unit of a filamentous fungi. A tubular filamentous structure surrounded by a cell wall and filled with cytoplasm, and growing from the tip.

Imperfect fungi – Fungi that seem unable usually to produce sexual fruit bodies were formerly placed in a category called imperfect fungi (deuteromycetes). With the advent of modern molecular techniques it is now possible to place these in the appropriate group in Kingdom Fungi.

Isidium (plural isidia) – An outgrowth from the thallus (body) of a lichen. An asexual reproductive structure.

Lichen – An organism formed from a combination of a fungus plus algal or cyanobacterial cells.

Lymph – A colourless body fluid containing white blood cells.

Macrofungi – Fungi that produce large sexual fruit bodies.

Mastigomycetes – This was previously a group of fungi which has now been split up into other groups including chytrids.

Microbe – A microscopic organism, such as a bacterium.

Microsporidia – Minute organisms now considered by many people to be fungi. They are obligate parasites within cells, particularly animal cells.

Mycelial cord – Cable-like structure consisting of an aggregate of closely interwoven, more or less parallel-aligned hyphae. Not growing from a definite apical growing point.

Mycelium – The main body of most fungi, made up of individual threads termed hyphae (singular hypha).

Mycobiont – The fungal partner in a lichen or mycorrhiza.

Mycoinsecticide – Some fungi that kill invertebrate pests can be grown up and made into a formulation that can be applied in the environment to kill appropriate insects. This avoids the use of toxic chemicals harmful to the environment.

Mycopesticide – See mycoinsecticide.

Mycorrhiza – A mutually beneficial association between fungi and plant roots in which the fungal hyphae (filaments) grow intimately with the root and extend into the soil. The fungus provides the plant with water and mineral nutrients, and the plant usually provides the fungus with sugars.

Mycotêtes – Literally fungus head, produced by the fungi cultivated in the nests of higher termites. The termites eat the mycotêtes as a source of food.

Necrotrophy – Obtaining nutrition by killing.

Nematicide – A pesticide that kills nematode worms.

Oomycota – A group of organisms that used to be classified as fungi, but are now clearly related more closely to brown algae and diatoms, and are no longer within Kingdom Fungi.

Organic resources – Material that can be used as a food source by fungi, such as plant, animal and microbial tissues.

Parasitism – An intimate relationship between one organism and an organism of a different species, in which one organism, the parasite, benefits nutritionally at the expense of the host.

Perithecia – Minute flask-shaped structures that contain the asci which contain the sexual spores of the ascomycetes called flask fungi.

Phagocytosis – A cell (white blood cell) or single-celled organism engulfs solid particles. It is a way in which some single-celled organisms feed.

Photobiont – The photosynthetic partner in a symbiotic relationship, such as algae or cyanobacteria in a lichen.

Photosynthesis – Plants combine carbon dioxide from the atmosphere with water using energy from sunlight to form sugars.

Rhizomorphs – Literally root morphology. A cable or cord-like mycelial structure formed by hyphal aggregation and growing from a clearly defined apex, often with a thick dark coloured outside rind.

Saprotrophy – Obtaining nutrition from dead organic matter.

Slime moulds – Fungus-like organisms whose reproductive structures superficially resemble those of true fungi.

Soredia – Powdery reproductive propagules produced by lichens. They are composed of fungal hyphae wrapped round green algae or cyanobacteria.

Spiracles – The breathing holes in the external surface of insects.

Spores – Any small dispersal, reproductive or survival units which separate from a hypha and can then grow independently into a new individual.

Stem cells – Unspecialised cells which are able to become specialised cells (e.g. a nerve cell in a vertebrate).

Symbiosis – Any more or less permanent intimate association between organisms. The relationship can be mutualistic or parasitic.

Thallus – The body of a fungus or lichen. The mycelium is the thallus or body of a fungus.

Toxin – A poisonous substance produced by a living cell.

Yeast – Unicellular fungi that multiply usually by budding, but also in some cases by splitting into two. Many fungi can have a yeast-like phase in their life cycle. Some fungi, for example brewers' yeast, spend all or most of their lives as yeasts.

Zygomycetes – A group of relatively simple fungi that do not usually have cross walls within their hyphae.

Zygomycota – See zygomycetes.

Dryad's saddle
(*Polyporus squamosus*).
Image: © Chris Jeffree.

Index